# FIRSTS in FLIGHT

Alexander Graham Bell and his Innovative Airplanes

Terrance W. MacDonald

Formac Publishing Company Limited
Halifax

Copyright © 2017 by Terrance W. MacDonald

All rights reserved. No part of this book may be reproduced or transmitted in any form or by any means, electronic or mechanical, including photocopying, or by any information storage or retrieval system, without permission in writing from the publisher.

Formac Publishing Company Limited recognizes the support of the Province of Nova Scotia through the Department of Communities, Culture and Heritage. We are pleased to work in partnership with the Province of Nova Scotia to develop and promote our cultural resources for all Nova Scotians. We acknowledge the support of the Canada Council for the Arts, which last year invested $153 million to bring the arts to Canadians throughout the country. This project has been made possible in part by the Government of Canada.

Cover design: Shabnam Safari
Cover image: Alexander Graham Bell National Historic Site

**Library and Archives Canada Cataloguing in Publication**

MacDonald, Terrance W., author
    Firsts in flight : Alexander Graham Bell and his innovative airplanes / Terrance W. MacDonald.

Includes bibliographical references and index.
ISBN 978-1-4595-0478-3 (softcover)

    1. Bell, Alexander Graham, 1847-1922.  2. Airplanes--Nova Scotia--Baddeck--Design and construction--History--20th century.  3. Airplanes--Nova Scotia--Baddeck--Flight testing--History--20th century.  I. Title.

TL523.M25 2017    629.13009716′93    C2017-903281-X

Formac Publishing Company Limited
5502 Atlantic Street
Halifax, Nova Scotia
B3H 1G4
www.formac.ca

Printed and bound in Canada.

*This book is dedicated to Autumn Lee Marie LeMoine, the beautiful baby girl whom I'm proud to call my granddaughter. Autumn received her angel's wings on April 16th, 2014. She gave us the best 5 months and 13 days of life a family could ask for.*

## Photo Credits

Bell family names and likenesses used with permission of The Alexander and Mabel Bell Legacy Foundation. All photos provided courtesy of the Alexander Graham Bell National Historic Site, except the following:
Author: 57, back cover (middle)
Aviation Knowledge (CCA): 33
Blair Fraser: 77, 78, 80
PilotFriend.org: 82 (top left)
Precision Graphics: 6
Wikimedia: 38 (bottom left), 51, 62
Wright-Brothers.org: 43

# Contents

| | |
|---|---|
| Introduction | 5 |
| **1** Discovery in a Self-Taught Flight School | 8 |
| **2** Bell's Team | 18 |
| **3** The First Planned Flight | 27 |
| **4** Evolution to Powered Flight | 32 |
| **5** Bell's Team Surpasses the Wrights | 41 |
| **6** Flying on Water | 47 |
| **7** What Happened in Baddeck? | 53 |
| **8** The Silver Dart, A Better Airplane | 59 |
| **9** The AEA Dissolves | 64 |
| **10** The Canadian Aerodrome Company | 68 |
| **11** The Baddeck No. 2 Aeroplane | 72 |
| **12** Unauthorized Passenger | 75 |
| **13** Advanced Thinking | 81 |
| **14** The Final Three Aeroplanes | 85 |
| **15** The Canadian Aerodrome Company Struggles to Survive | 88 |
| Endnotes | 93 |
| Additional Sources | 95 |
| Acknowledgements | 95 |
| Index | 96 |

# Introduction

The biggest name in Canadian flight history is Alexander Graham Bell, inventor of the telephone. Over one hundred years ago, when aviation was still in its infancy, Bell was able to think like today's highly trained and experienced pilots. His work in aviation was nothing short of brilliant, particularly when we consider this was a man who never flew once, not even as a passenger. Despite his brilliance, Bell tasted failure more often in the study of flight than in any of his other pursuits — an indication of the sheer magnitude of challenges that all pioneering aviators faced in developing a safe flying machine.

This story shines a spotlight on the numerous, yet lesser known, aviation experiments that occurred in the quiet, peaceful community of Baddeck, where Bell spent decades of his aviation career. Mabel Bell, his wife, always felt he never received the credit he deserved for his involvement in aviation. Wanting her husband's work documented and never forgotten, she said, "I feel that the Alexander Graham Bell that I see must be very different from the Alexander Graham Bell outsiders see . . . If a regular biography is ever written, it should be by one at greater distance . . . My feelings would be to have someone want to write it from his own feelings of fascination . . . do it for his own pleasure."[1]

Bell leans over to hear Mabel as they sit in the motorboat, Ranzo, during a conversation at the dock just below their lodge.

This is that aviation story.

Bell's telephone fame shouldn't overshadow the other amazing things he accomplished in his lifetime, including in flight. In order to fully appreciate Bell's aviation accomplishments in his late sixties, one has to know his background, what he achieved at an early age and how he developed his various inventions. His tenacity, independent thinking, practice of working with others and attention to detail — all integral to his success with the telephone — were critical to his achievements with aircraft. His story is fascinating, and while Bell was never officially a Canadian citizen, there are many significant aspects of his story that took place in Baddeck.

Bell spent his last thirty years increasingly at Beinn Bhreagh (pronounced "ben VREE-uh"),[2] Gaelic for "beautiful mountain," sometimes living there year-round. The years spent there did not signal slowing down or retiring, as Bell declared, "Wherever you may find the inventor, you may give him wealth or you may take away from him all that he has; and he will go on inventing. He can no more help inventing than he can help thinking or breathing."[3] He once said, "I want many more years of life to finish it all. I cannot hope to work out half the problems in which I am interested."[4]

The early 1900s was a very dynamic period, comparable to the space exploration and computer development periods that occurred later in the twentieth century. In fact, a mere sixty years separated American astronaut Neil Armstrong's first steps on the moon from the first flights of Douglas McCurdy in Canada.

*Beinn Bhreagh Hall seen from the Bras d'Or Lake with Bell's tetrahedral tower at the top of the mountain on the right.*

In 1909, the residents of the small town of Baddeck had no idea that powered flight would eventually lead to people walking on the moon. Once it began, engine-powered flight developed rapidly in North America; however, the significant Canadian contributions from the flying machines developed by Bell and his team are often overlooked in the history of heavier-than-air powered flight.

Depending on your knowledge of aviation history, Canada's first airplane may not be the one you thought it was, and the first Canadian pilot to fly probably isn't who you think it is, either. Bell was a great aviation pioneer and the Baddeck aviation story consists of much more than just the Silver Dart. The aviation team that Mr. and Mrs. Bell put together was recognized as stiff competition to the Wrights. This story will introduce each member of that team, including each person's strengths and skills before joining Bell, their valuable contributions to the team during its short existence and each person's work after the team dissolved.

Most Canadians don't know the full Baddeck aviation story, or the influence that Bell's work had on the development of flying machines generally. This book outlines the major aviation events and important ways that Bell and his team contributed to the advancement of aviation in Canada, and the world.

# 1 DISCOVERY IN A SELF-TAUGHT FLIGHT SCHOOL

On the cold afternoon of February 23, 1909, long before there were any airports, the frozen surface of Baddeck Bay was Mother Nature's perfect runway for the historic flight attempt. The doubtful people of Baddeck were uttering dire predictions for the "dubious" adventure that was about to occur. Without a seatbelt or helmet, and with eight cylinders of a concept motorcycle engine thundering behind him, self-taught pilot Douglas McCurdy opened up the throttle and started the takeoff run across the ice. Sitting on a hard wooden plank seat surrounded by guy wires, sticks and the canvas that made up the flying machine named the Silver Dart, McCurdy proceeded to lift the shining bird into the sky above Baddeck.

That's the story most commonly told of how Bell's aviation career began. However, his interest in flight started long before that day and continued for years afterward. He lived from 1847 to 1922, and in his lifetime he and his team of engineers improved airplane design enough to come close to today's standards.

Bell patented his best-known invention, the telephone, in 1876 at the age of twenty-nine; because this was the single most valuable patent ever awarded, Bell earned enough that he had no need to pursue commercial applications. Being a good businessman and managing money weren't among his strengths, but his wife, Mabel, with her common sense and business acumen, helped him manage. Free to pursue his many interests, Bell held more than thirty patents — eighteen in his name alone, plus twelve that he shared with collaborators — with his contributions to aeronautics reflected in the nine patents issued for various advances in flying machines. The US Patent Office declared Bell first on its list of the country's greatest inventors (1936).

In his birthplace of Edinburgh, Scotland, Bell devised his first practical invention in 1858. He and a friend, Benjamin Herdman, were playing around the Herdman flour mill when Ben's father admonished them to do something useful. When Alexander asked what they could do, John Herdman held out some grain and told them they would be a big help if they could remove the husks. Alexander surprised him when he produced an apparatus to do just that, by attaching wire brushes to a set of rotating paddles on an existing machine.

Bell was, by his own admission, a poor student. A brooding loner, he confessed to a propensity to dream, stimulated by the love of reading instilled in him by his grandfather. Dreaming, botanical collecting and rambling at Milton Cottage in Buckinghamshire, England, fuelled Bell's curiosity and his desire to invent. The darkroom installed by Melville Bell at Milton Cottage sparked in Alexander a keen interest in the new artform of photography, which he would use to document his work throughout his life.

A pivotal event in Bell's life came when he was sent to London to spend a year with his recently widowed

*Bell and daughter Elsie at the Bell family home in Brantford, Ontario.*

grandfather. Thinking the boy rather indolent, the elder Alexander set out to make certain that he learned Latin and became well read. Alexander was a loner without many companions his own age, so he attended his grandfather's speech classes. He looked back on the time spent with his grandfather as a turning point, when he reached the important decision that teaching would become his life's work.

After tuberculosis took two of his brothers, the Bell family immigrated to Canada; after arriving in Quebec on August 1, 1870, they travelled to Brantford, Ontario, where they bought a home in Tutelo Heights.

To the very end of his life, Bell maintained the pure delight of a child exploring the world, while those who knew and loved him worried about his lack of concentration. Bell was a great generalist during the age of specialists, and only five months before the telephone was patented, his future father-in-law chided him about his inclination "to undertake every new thing that interests you and accomplish nothing of value to any one."[5]

While walking along a beach in Scotland, Bell scribbled notes on bird activity and paid particular attention to their tails. Inspired by a seagull in flight, he drew a flying machine complete with ailerons. He titled the notes "Aerial Aviation," and sketched a crude design of a flying machine that bears a striking similarity to drawings made by the Renaissance artist and inventor Leonardo da Vinci. (Scholars who have examined Bell's notebooks are often struck by the parallel between these two great generalists.)

Aeronautical science was not a sudden whim of Bell's, as Thomas A. Watson, his aide in the invention of the telephone, recalled: "From my earliest association with Bell, he discussed with me the possibility of making a machine that would fly like a bird. He took every opportunity that presented itself to study birds, living or dead . . . I fancy, if Bell had been in easy financial circumstances, he might have dropped his telegraph experiments and gone into flying machines at that time."[6] Bell enjoyed and preferred the challenge of aviation that much.

As a boy, he would lie atop a favourite hill in Edinburgh, Scotland, so he could be closer to the sky and watched with envy and wonder as the birds flew above him. All through his youth, he continued to study the soaring of birds and constantly tinkered with wings and propellers. On his honeymoon in September 1877, he told his wife, Mabel, that he dreamed of flying machines with telephones attached.

The accidental deaths of aviation pioneers tempered Bell's enthusiasm for manned flight with a concern for safety. Being stable and safe in the air was the overriding interest in all his aeronautical experiments. "How," he asked himself in a notebook, "can ideas be tested without actually going into the air and risking one's life on what may be an erroneous judgment?"[7] Bell was enticed by the notion of using flying kites in many of his aviation tests, because he felt that the kites permitted important initial in-flight experiments without risking human life.

Although Bell had become a naturalized American citizen in 1882 after his move from Scotland, his previous ties to Canada (his time in Ontario from August 1870 to April 1871) were strengthened again in the summer of 1885, when he and Mabel visited Baddeck, on the Bras d'Or Lakes of Cape Breton Island. They were drawn by Charles Dudley Warner's book on Baddeck, and by the temperate climate — Bell hated hot summer weather. To his delight, Bell found that the Baddeck area and its people reminded him of Scotland. The following summer they rented a cottage (which they later bought) and began buying up land. For the next thirty-six years, until he died, Bell would divide his time between Washington and Cape Breton.

Much of his scientific work was done at Beinn Bhreagh, Bell's six-hundred-acre estate constructed on a peninsula — known as "Red Head," due to the reddish sandstone rocks at its tip — that juts into Cape Breton Island's scenic Bras d'Or Lake. Located approximately three kilometres southeast of the village of Baddeck, it forms the southeastern shore of Baddeck Bay. When they were in residence at Beinn Bhreagh from spring to fall, the Bells were active members of Baddeck society, and the thirty-seven-room mansion always seemed to vibrate with family and guests. They came to know Cape Breton intimately.

Throughout the 1890s, residents of Baddeck were accustomed to looking up at Beinn Bhreagh and seeing the red kites flying in the sun. No doubt, many of these practical people whispered among themselves as they watched Bell's designs waft over the Bras d'Or Lakes.

Bell and his associates would conduct more than 1,200 carefully documented, flight-related experiments over twenty years, most of them in Baddeck. To the residents of Baddeck who watched, they had a fanciful air. A boatman who observed one flight stopped short, in an account cited by John Hamilton Parkin, of describing Bell as a lunatic: "He goes up there on the side of the hill on sunny afternoons and with a lot of thing-ma-jigs and fools away the whole blessed day, flying kites. Mind you, he sets up a blackboard and puts down figures about these kites and queer machines he keeps bobbing around in the sky. Dozens of them he has . . . It's the greatest foolishness I ever did see."[8] Nothing better illustrates Bell's independence of thought than his staunch support of aviation at a time when it was considered so quixotic a subject that Bell risked his scientific reputation in pursuing it.

As his kite structures became larger, he required staff to assist with their construction, and eventually a large building was erected, in which all the work could be done and where the kites could be stored between flights. The kite house was a long, narrow structure without partitions, having one side wall that opened on rollers to let a kite escape when finished. In spite of

*Bell flying a tetrahedral kite in the kite field at Beinn Bhreagh in 1907.*

some residents' doubts, the population of Baddeck was soon pulled into Bell's kite flying activities — young girls sat sewing in the kite house, surrounded by piles of bright red silk, while the young men were engaged to take photographs or work the pulleys.

In 1898, Bell built Jumbo, a box kite similar to Lawrence Hargrave's, which he thought was of very sound design. This craft, measuring fifteen feet long, almost eleven feet wide and five feet deep, was big enough to carry a person. Bell used bigger boxes than Hargrave, with the belief that he would be able to achieve a more powerful and stable lift. Unfortunately, Jumbo was too heavy and never flew,

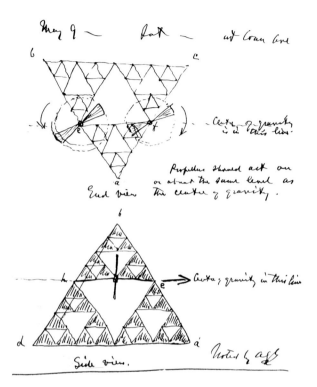

Bell's notes for May 9, 1909, showing a powered tetrahedral kite, his version of the ultralight trike, which does exist today.

but it was one of Bell's first attempts to build a man-carrying kite. Bell persevered, though — his concern was always about stability rather than speed as he experimented with different sizes of box kites.

Looking for a stronger and lighter structure, he began combining and arranging triangles. Knowing there is more strength in a triangle compared to a box, in his fifty-fifth year Bell discovered the geometric form known as the tetrahedron, a triangular pyramid. Bell planned on using this structure to build a kite-like airplane. The tetrahedral kite was put into mass production at the kite house, and thousands of individual cells were built in various kite configurations to test his aviation theories. The tetrahedral kite, while not easy to build compared to other kites, was very stable, and it flew well in moderate to heavy winds if properly set up.

Bell's idea was that it was necessary to compound the kites, removing the extra stick where the supports came together — these kites had the same lifting power as the individual kites, but with less weight. Multicelled kites allowed Bell to explore this concept; the tetrahedral shape was stronger and less resistant to wind than a rectangle, and in his notebook entry for August 25, 1902, Bell recorded the excitement he felt as he linked the perceptions he'd gleaned from months of tests. Frustrated that he could not draw the cell well, he described the shape as

> *a figure composed of 4 equilateral triangles having 4 triangular faces bounded by 6 equal edges. Wish I could describe this solid form properly, as I believe it will prove of importance not only in kite architecture — but in forming all sorts of skeleton frameworks for all sorts of constructing — a new method of architecture.*[9]

The summer and fall of 1902 found the laboratory at Beinn Bhreagh experiencing its greatest period of activity yet, as the construction of the larger tetrahedral kite space frames began and work was underway on the frame of the

Cygnet I kite. Huge tetrahedral kites were suspended from rafters during construction.

Bell's notes from November 18, 1902, read:

> *Two great successes today, both a result of suggestions from Mabel. First suggestion. Instead of waiting for wind, attach kite to galloping horse. Tried it yesterday with small kites. So promising that we tried three of our large kites today in same way. Found I could study their mode of flight in the air as well as if I had wind . . . or nearly so . . . and could judge of their way of falling better than with wind.*[10]

Bell felt this method would be a better way of judging a landing because the horses could be slowed down gradually, which is how an airplane lands, by slowing down and descending closer to the earth. Mabel's second suggestion involved hairpins and sealing wax for linking the tetrahedral skeletons, which Bell exclaimed was "just the thing."[11]

Bell was also learning about takeoffs and what pilots now call the "angle of climb"[12] for their aircraft, which is the angle at which Bell needed to set his kite to become airborne against the relative wind generated by the forward movements of the horses or by a natural wind. Once, when a five-year-old boy was trying to fly his toy kite, Bell, who'd been passing by, said, "That kite isn't bridled properly."[13] He pulled it down and re-bridled it, and, sure enough, when he launched it again it flew better.

Bell's tetrahedral experiments spawned a cottage industry in Cape Breton, as hundreds of farm homes

*Workman Hector McNeil stands beside a tetrahedron frame, one of the triangular pyramid pieces that made up Bell's Cygnet I.*

helped make cell frames of spruce, bamboo and metal. In October of 1903 Bell switched his spar (kite framing) material from black spruce to strong but light aluminum tubing. Suddenly, these large kites with fanciful names became much more buoyant and were manoeuvered about with greater ease, flying in a lot less wind. In Baddeck, Bell continued to experiment with the tetrahedron, which he patented on August 31, 1904.

In addition to the Cygnet I, Bell built a lesser known kite made of 1,300 tetrahedron red silk cells, capable of lifting a person off the ground. In December 1905, the kite was lofted in celebration of a wedding at the Beinn Bhreagh estate. Arthur Williams McCurdy — a native of Baddeck, editor of the local newspaper and Douglas McCurdy's father — was

*Workers run to remain clear of the descending Frost King kite during test flights in the kite field at Beinn Bhreagh.*

Bell's secretary, photographer and virtually a member of the Bell family. When McCurdy's daughter, Susan, was betrothed to Walter Frost, the Bells insisted that the nuptials be performed in their big house. As soon as the couple were pronounced married, Bell christened his new kite the Frost King in their honour.

Each cell weighed under an ounce and the kite's total weight was only sixty pounds. On December 18, 1905, the kite supported Lucian McCurdy (Arthur McCurdy's brother) on a rope, and in doing so it raised four times its weight — about 243 pounds. Bell then equipped the kite with a fifteen-foot rope ladder, which would hang by lines some thirty feet below the kite itself. The plan was for a lightweight man to climb the ladder when the kite was in flight, then ride it higher into the air to prove the kite's lifting power. The rider would not control the kite, but the sheer act of lifting a person well off the ground was a key step. On Christmas Day, a good breeze lifted the kite some nine hundred feet in the air.

On December 28, 1905, the Bells were due to catch a steamer boat bound for Saint John, New Brunswick, where they would have boarded a train for Washington. However, Bell was so dedicated to his study of flight that he elected to take advantage of the "fair sailing breeze" that had come up, offering one last chance to fly the Frost King. The kite lifted its handler (Neil McDermid) thirty feet above the ground. The kite, including all of its tackle, weighed just 125 pounds; the weight of the handler was 165 pounds, while a ten-mile-per-hour wind was blowing. The pull of the kites was measured using a standard spring scale, and from this data Bell concluded that a much larger kite, carrying an engine that provided a ten-mile-per-hour thrust, would easily carry a person.

## The Worldwide Pioneering Aviation Community

The fifty-year period around the second half of the nineteenth century was a time of great inventions. Between 1860 and 1910, the forerunners of many objects we use today were first produced, such as cameras, records, movies, typewriters, telephones and radios. The planet came to life with electric lights, automobiles and airplanes. However, it was the intelligence, courage and strong desire of pioneering inventors and aviators who accomplished this; without them, humans would still be restricted to travelling around the earth in automobiles and boats.

The aviation community during this time was a very small group of individuals spread all over the world. In spite of distance and slower modes of communication, this small group was close-knit. They not only competed with each other, but willingly shared information on the advancement of flight.

The linchpin that united and kept this group in close communication was Octave Chanute.

Chanute, of Chicago, Illinois, was an American civil engineer and aviation pioneer. Born in France, he immigrated to the US in 1838. When he retired from his railroad career in 1883, he decided to devote some leisure time to furthering the new science of aviation. Chanute collected all available data from flight experimenters around the world and combined it with his knowledge gathered as a civil engineer. He published his findings in the influential 1894 book *Progress in Flying Machines*. This was the most systematic global survey of fixed-wing, heavier-than-air aviation research published up to this time. Chanute became an independent engineering consultant, providing advice to many budding aviation enthusiasts and helping to publicize their flying experiments. At his death he was hailed as the father of aviation and the heavier-than-air flying machine.

Chanute corresponded with many aviation pioneers, including Bell, Louis Bleriot, Albert Santos Dumont, Ferdinand Ferber, Lawrence Hargrave, Samuel Pierpont Langley, Otto Lilienthal, John J. Montgomery, Louis Mouillard, Percy Pilcher, Gabriel Voisin and the Wright brothers. Bell himself was known to be in contact with Hargrave, the Wright brothers and Langley.

On the latter, Bell once said, "For many years past, in fact from my boyhood, the subject of aerial flight has had a great fascination for me and I was therefore much interested in the research of Langley."[14] On June 15, 1891, Bell was invited by Langley to witness

*Men flying the Frost King tetrahedral kite in Baddeck in December, 1905.*

some of his model flying machines as he ran them through experimental flights. Bell witnessed Langley's large, steam-powered, model aeroplane experiments and wrote, "The sight of Langley's steam aerodrome circling in the sky convinced me that the age of the flying machine was at hand."[15]

Bell became good friends with Langley, a fellow Washingtonian, and he found it impressive that such a respected scientist would be engaged in the study of

*Langley's Aerodrome Number 5, an unpiloted, steam-driven aeroplane, flew for three-quarters of a mile over open water on May 6, 1896.*

flight with the goal of creating a machine that could fly. In June 1892, Bell dictated an article entitled "The Flying-Machine of the Future,"[16] in which he described passenger-carrying machines capable of long-distance flights that he believed would be possible once some of the problems of flight were solved.

Bell worked with Langley on things like gunpowder rockets and the rotating blades of helicopters. Langley attempted to make a working, piloted, heavier-than-air aircraft, but his two attempts at piloted flight were not successful. Langley made larger flying models powered by miniature steam engines — petrol had not yet arrived — which were a marvel of delicate craftsmanship. His first success came when his Number 5 unpiloted model flew nearly three quarters of a mile after a catapult launch from a boat on the Potomac River, photographed by Bell. The distance was ten times longer than any previous experiment with a heavier-than-air aircraft, demonstrating that stability and sufficient lift could be achieved in such craft. On November 11 of that year, his Number 6 model flew more than 5,000 feet (1,500 metres).

The greatest problem confronting Bell at this time was how a heavy engine, let alone a person, could be supported by a machine in the air. Bell was aware that Hargrave had lifted himself off of the ground under a train of four box kites in Australia, rising sixteen feet in a wind speed of twenty-one miles per hour. This experiment established the box kite as a stable aerial platform, which greatly interested Bell, and he travelled all the way to Sydney, Australia, just to meet Hargrave and discuss flight.

As Bell wrote on September 2, 1901, "The great difficulty in developing an art of aerial locomotion lies . . . in the difficulty of profiting by past experience . . . a dead man tells no tales."[17] Bell held continued discussions with Langley, both in Washington and at Beinn Bhreagh, when Langley visited for an extended period in the summer of 1903. These extended conversations fuelled Bell's desire to know more about flight, and again

Langley invited Bell to witness experiments with his largest machine, designed to carry a person. Bell was present at the spectacular crashes into the Potomac River of Langley's full-sized, manned glider on October 7 and December 8, 1903. Designated pilot Charles Manly was unhurt in both accidents; however, since Langley's work had been funded by the US government, the press treated Langley's very public experimental failures in flight with much derision. Langley was devastated by the public humiliation. Pioneering aviators tried to avoid ridicule by practising out of the sight of neighbours and the general public. For example, in 1804, George Cayley, a British aeronautical engineer, tested a glider at night to avoid attention. The treatment of Langley by the press had a profound effect on Bell and from that time forward he proceeded with even greater caution in his work, recognizing that published opinion could tarnish reputations and slow innovation.

When Mabel first suggested that Bell should have help, he explained to her that he was not ready yet. He said,

> *You must remember that this is not a question of invention but discovery — and discovery is groping if you will — a slow laborious systematic groping after knowledge . . . disheartening in the number of blind alleys explored — and yet this process of groping carefully and systematically all round, in every direction, must lead at last to full knowledge and the discovery of the true path . . . I want too, to bring my experiments to a*

*Alexander Graham Bell monitoring the flight of a large tetrahedral kite over the kite field at Beinn Bhreagh.*

> *conclusion — and can do so here — if let alone. Then will come the work of invention and then will be the time when I can follow your suggestion — of having half a dozen men or more in my laboratory instead of two. But now — I could not tell them what to do.*[18]

When it was time, and when he was confident enough in what he was learning from kites, he would seek some helpful colleagues.

That day came. The immense size of the kites soon dwarfed the design expertise of Bell and his staff at Beinn Bhreagh, and Bell believed that the substitution of an engine and a propeller attached to the kite might permit free person-carrying flights, so in the summer of 1906, Bell and Mabel set out to recruit some younger talent. Bell felt they should not limit themselves to the tetrahedral construction alone, but endeavour, by any means possible, to get a person into the air.

# 2 BELL'S TEAM

Lawrence Hargrave wrote in 1893:

> *The flying machine of the future will not be born fully fledged and capable of a flight for 1000 miles or so. Like everything else it must be evolved gradually. The first difficulty is to get a thing that will fly at all. When this is made, a full description should be published to aid others. Excellence of design and workmanship will always defy competition.*[19]

During the period from 1900 to 1911, the pioneering groups building the first airplanes were in competition with each other as they progressed through the development of kites and then planes, yet they absorbed and incorporated each others' progress. No one believed at the time that just a few short years later the airplane could play such an important role in World War I.

After experiencing some success on his own, Bell recruited four young men, and together they formed a scientific team incorporated as the Aerial Experiment Association (AEA).

The AEA would not have been possible without Mabel's contributions. While Bell was busy with his scientific enterprises, he had complete confidence in

Mabel to efficiently manage the family finances and the operation of their two homes in Washington and Baddeck. An avid supporter of Bell's work, Mabel had urged him to take on some associates to help him. Being independently wealthy, Mabel provided a total of US$35,000 to finance the association, with $20,000 available immediately. Mrs. Bell, the silent sixth member, sold a house lot she owned in Washington to fund the operation. Mabel's sensible suggestion made her the first woman in history to propose, establish and fund a research group.

Bell and Mabel initially approached John Alexander Douglas McCurdy, a fearless native of Baddeck and son of Arthur Williams McCurdy, Bell's chief assistant at Beinn Bhreagh. At Mabel's suggestion, Douglas encouraged Frederick Walker "Casey" Baldwin, a university friend, to visit Baddeck and meet the famous Bell. Thomas Etholen Selfridge, a twenty-five-year-old lieutenant in the United States army and the military's authority on aviation, contacted Bell to find out more about his experiments, and through Selfridge, Glenn Curtiss was invited to take part in Bell's group to develop a flying machine.

Bell thought the young men of the new association should be given some personal inducement to continue cooperating together. Such inducements might be for each man to make a name and reputation for himself as a promoter of aviation or mechanical flight, and, in the event of any profits, for each to have a share. But the chief inducement that would keep them together would be the actual accomplishment of aerial flight and the honour and glory that would attach to those who succeeded.

*Mabel Bell, always active in Bell's experiments, holds a device for measuring kite "pull."*

Bell observed that his youthful friends probably looked upon the money offered by Mrs. Bell as the total capital involved without knowing how much he had already spent in his prior aviation experiments, which he estimated at $100,000 over a period of fifteen years. He felt that the young men came in not at the beginning, but near the conclusion of a long series of experiments to which they had contributed nothing. Bell insisted that a few veteran employees, who had helped him develop the kites, deserved to be members of the association. When the young men objected to this, Bell was irked. "This is the first snag I have struck, but there are others."[20] He emerged with a fresh formula: "The past belongs to the past and the future to the future — the Association to deal with only the future."[21]

The AEA's charter was ratified in Halifax, Nova Scotia, on September 30, but it didn't take effect until the next day, October 1, 1907. Bell arranged to have the signing with the notary public witnessed by the US Consul to Nova Scotia, Mr. David F. Wilder. They had planned the Halifax ratification after Glenn Curtiss had advised his colleagues of Captain Thomas Baldwin's (no relation to Casey) forthcoming booking at Halifax in the first week of October, where he had an engagement for dirigible flights. The entire group, Mabel included, boarded a train for Halifax. At the earliest opportunity after the official signing of the AEA, the Baddeck delegation went out to the Halifax fairgrounds, where Captain Tom Baldwin, an American balloonist of significant accomplishment, performed his dirigible flight. Seeing the first-ever flight conducted over Halifax, Bell expressed himself as "delighted."

The six associates of the AEA came from two distinct generations and three nationalities. Bell was an American citizen with strong Canadian roots and an early childhood spent in Scotland. Mabel, the American wife of Bell and very prominent in the affairs of the AEA, is referred to as Bell's "silent partner"; her total contributions to the AEA are now more widely recognized. McCurdy and Baldwin were Canadian, while Curtiss and Selfridge were American.

Bell pointed out to his AEA team that, based on his experience with the telephone, they should not hope for great financial returns from the construction of flying machines, as litigation over patent rights would destroy profit for many years.

Bell referred to Douglas McCurdy as "a doer." After some education in the village school at Baddeck, McCurdy was sent to Ontario to complete his education. The Bell family assisted with paying for McCurdy's education in mechanical engineering at the University of Toronto, from which he graduated in 1907. Following graduation, McCurdy returned to Baddeck to work with Bell and signed on as a full member of the AEA for the duration of its existence.

While at university, McCurdy became good friends with Casey Baldwin. In 1906, Mabel suggested to McCurdy that during the summer break he bring one of his fellow university students back to Baddeck for a couple of weeks — someone who might be interested in what her husband was doing. As it turned out, one of the most significant events in 1906 was the recruitment of Baldwin, a graduating engineer from the University of Toronto.

Baldwin was born in Toronto, Ontario, the grandson of Sir Robert Baldwin, a Canadian prime minister prior to Confederation. Baldwin attended the University of Toronto, where he studied both electrical and mechanical engineering. He graduated in the spring of 1906, a full year ahead of McCurdy, and travelled to Baddeck with McCurdy that summer. Bell, aged fifty-nine, was impressed with twenty-four-year-old Baldwin's enthusiasm for his

*The members of the Aerial Experiment Association at Beinn Bhreagh (left to right): Glenn Curtiss, Casey Baldwin, Alexander Graham Bell, Lieutenant Thomas Selfridge, Douglas McCurdy.*

aerial experiments and invited him to stay, referring to him as "a thinker." Baldwin accepted Bell's invitation, and the couple of weeks he intended to stay turned into forty satisfying years — literally the rest of his life. Baldwin became a working partner of Bell's and a virtual extension of the elder inventor. The two men worked very closely on everything and had complementary skillsets.

Baldwin's attitude toward Bell was not that of a paid employee, but as a graduate of a technical school. Baldwin worked with Bell through the fall of 1906 and stayed on through the winter and spring of 1907, while McCurdy returned to university. Baldwin gathered information on the works of Lilienthal, Chanute, Langley and the Wright brothers to expand Bell's library of materials on the topic of aviation. The two men built many tetrahedral models to test their theories of kite construction. Aside from Bell, Baldwin was the brains behind the AEA, making him the logical choice to be chief engineer. McCurdy and Baldwin towed kites with motorboats when the breeze was right, measured wind velocities and

*AEA Chief Engineer Baldwin behind the controls of the White Wing on May 18, 1908, at Hammondsport, New York.*

altitudes, tested engines and propellers and had a host of other experimental duties.

As historian John Boileau points out, "Chance brought Bell to Baddeck, and chance intervened again to bring Baldwin there; in both cases, it was an extremely productive relationship."23

J.H. Parkin wrote,

> Bell, the elderly scientist, and Baldwin, the young engineer, complemented each other to constitute a singularly effective team. Baldwin acted as a foil to Bell's active brain, and his sound, down to earth, practical approach served to bring some of Bell's more exuberant proposals within range of achievement. He converted the ideas of Bell into working devices. Despite Bell's efforts to push him to the center of the stage, Baldwin was content to remain in the wings.

> *In consequence the real Baldwin is virtually unknown . . . Those who knew him personally were aware of his sterling qualities — cool, resourceful, unselfish, a most agreeable companion, and staunch friend. He had the same regard for truth as Bell; indeed it was a bond between them. Certainly theirs was an association of a rare kind, and from it and in it came the dawn of Canadian aviation.*24

John Boileau notes, "With Baldwin at his side, Bell proceeded to construct an 'aerodrome,' his preferred term for a flying machine. Bell maintained that an 'aeroplane' was merely a wing or lifting surface, while an 'aerodrome' was an entire machine."25

The word aerodrome, coined from Greek words roughly translated as "air runner," was an early aviation expression used by Langley before Bell. Both men numbered all their successive models, and this is why the Silver Dart is sometimes referred to as Aerodrome Number 4. To make use of present terminology, going forward the aerodrome shall be referred to here as an airplane and hydrodrome shall be referred to as a hydroplane.

In the fall of 1907, two American visitors arrived at Beinn Bhreagh who changed the direction of the team's exploration in the science of flight for the better: Thomas Selfridge and Glenn Curtiss. Thomas Etholen Selfridge was born in San Francisco, California, and he attended the United States Military Academy at West Point and graduated in the top third of his class. After graduation, he was stationed in Fort Myer, Virginia, near Washington.

Having developed an interest in aviation while at West Point, Selfridge had read all he could on the fledgling efforts to successfully build and pilot airplanes. The US Army was looking at the military potential of manned aircraft and had purchased a dirigible, Army Dirigible Number 1, in July 1908, from Captain Thomas Scott Baldwin.

Lieutenant Selfridge strongly believed in the potential of aviation as an effective tool for army operations. As a result, in 1907 Selfridge approached the Wright brothers to voluntarily assist them in their work on airplane design. The secretive Wright brothers declined Selfridge in a letter, stating that they had no need for part-time staff and that they were not adding any full-time people to their operation. This rejection set the stage for a relationship between Lieutenant Selfridge and Bell's team.

Lieutenant Selfridge had read about Bell in *National Geographic*. Being stationed near Washington, Selfridge had occasion to hear Bell speak on aviation and his experiments with kites in an address to the Smithsonian. He approached Bell and had a conversation with him that thoroughly impressed the scientist and made a lasting impression on him. Bell was a friend of President Theodore Roosevelt and so he wrote to him to have the interesting young officer officially detailed to Baddeck. Two days after its formation on August 1, 1907, Selfridge was assigned to the aeronautical division of the US Signal Corps, and shortly thereafter he received orders to proceed to Baddeck for the study of aeronautics in general and of heavier-than-air craft. As the US Army's official observer, Selfridge

*Lieutenant Selfridge at the controls of the White Wing on May 28, 1908, in Hammondsport, New York.*

arrived at Beinn Bhreagh in September of 1907 to join Bell, Baldwin and McCurdy.

Glenn Hammond Curtiss was born in Hammondsport, New York, a small town at the southern end of Lake Keuka. After one of his motorcycles found its way to Captain Thomas Baldwin, in 1904 Curtiss became a supplier of engines for the semi-rigid balloons that the captain was pioneering. That same year, Baldwin's California Arrow, powered by a Curtiss 9 HP V-twin motorcycle engine, became the first successful dirigible in America.

While exhibiting his motorcycles at the New York Automobile Show in 1906, Curtiss came into contact with Bell, who had been invited to exhibit his large tetrahedral kites. Bell, who was looking for a light, powerful motor to lift his manned kites, was very impressed with Curtiss. Curtiss was invited to Beinn

*Glenn Curtiss tests an engine and propeller on a boat in Baddeck.*

*On September 8, 1906, in front of the Beinn Bhreagh shoreline hangar, three of Bell's workers have the Ugly Duckling barge configured for propeller testing on Baddeck Bay.*

Bhreagh to see Bell's kites and discuss the potential of using engines to provide power for them.

Curtiss arrived at the same time his friend Lieutenant Selfridge was also visiting Bell to discuss aviation matters. The young men and Bell hit it off very well and spent days with the kites, trying various experiments to learn more about their lift. One of their first projects together was the Ugly Duckling, a catamaran-style boat designed to tow kites out onto Baddeck Bay. The young men spent evenings socializing with the Bell family and having energetic conversations on flight for hours after dinner in the grand house. Mabel Bell noticed the synergy of the young men, who added energy and focus to Bell's work. Glenn Curtiss helped the AEA to construct a proper engine; however, he did not share the comradeship enjoyed by the other three young men. Due to Curtiss's commitment to his motorcycle manufacturing plant in Hammondsport, he was preoccupied with his main business and as a result didn't put in the same amount of time or share the same enthusiasm as his AEA comrades. For Curtiss, it was more about business than friendship.

Before joining the AEA, Curtiss had written to the Wright brothers on May 16, 1906, and suggested he sell them one of his motors. The Wrights met with Curtiss three months later, when Curtiss travelled to Dayton while on a longer business trip. Although the meeting was cordial, the Wrights declined his offer and continued to build their own engines in secrecy using their own mechanic. In the fall of 1907, Glenn

Curtiss decided to join the AEA team, which was his second choice. Three months after he joined the AEA, though, on December 30, 1907, Curtiss once again offered the Wright brothers one of his new engines. Again, the Wrights declined his offer and kept him out of their camp. Hoping to be the first to cash in on the invention of the airplane, the Wrights viewed Curtiss as a threat to their success. The AEA was a non-profit association that didn't view Curtiss in the same way. The AEA offered the motorcycle engine businessman his best and only opportunity to learn about the theory of flight. Curtiss did deliver a valuable contribution to the AEA's flight attempts — without a good engine, AEA flights would never have happened. Bell regarded Curtiss as the leading manufacturer of motorcycle engines in the US.

The Beinn Bhreagh estate was designated as headquarters for the AEA, with a proviso: On or before January 1, 1908, the association would move to the US because of the rugged Nova Scotia winters, where work could continue in a more temperate latitude, after which a return to Baddeck was envisioned. Bell wanted to have his Cygnet kite flown before moving south.

Bell's laboratory was used without charge. He would also serve as chairman without salary; Baldwin was chief engineer, and he and McCurdy, as treasurer and assistant engineer, would each get $1,000 a year. Curtiss, who was the director of experiments, got $5,000 a year when on the scene and half that amount when away, as they realized that Curtiss's business commitments would not permit him to give the AEA his full time. Selfridge

*Glenn Curtiss, an American motorcycle manufacturer and member of Bell's team, in Hammondsport, New York, on July 15, 1907, with an engine he later brought to Baddeck.*

was secretary for the association; as he was already on full pay as an army officer, he honourably declined a salary.

Bell himself once said:

*For many years I have been a student of what is being accomplished in relation to aerial flight and for many years my laboratory experiments have been directed mainly along lines leading up to aerial flight as a logical conclusion. I now have associated with me four gentlemen who supplement my deficiencies with their technical knowledge. In this combination I now feel, united we are strong, whereas before we individually were weak.*[26]

Rannie Gillis of the *Cape Breton Post* wrote,

*The A.E.A. was now indeed a stronger unit. It is hard to imagine that Alexander Graham Bell had any deficiencies, yet back in 1907 he was honest enough to admit he would need some expert engineering help if he was to build a successful flying machine. He might have been one of the world's greatest scientific inventors, yet at the age of 60 he also realized he was too old to actually take part in any of the dangerous test flights that would be required in order to achieve success.*[27]

Bell described the fortuitous coming together of the group:

*We breathed an atmosphere of aviation from morning till night and almost from night to morning. Each felt the stimulation of the discussion with the others, and each developed ideas of his own upon the subject of Aviation, which were discussed by all. I may say for myself that this Association with these young men proved to be one of the happiest times of my life. I think that the progress of the experiments will be greatly promoted and the world benefited if these young men who are temporarily associated with me can be given some personal inducement to continue to cooperate together in accomplishing the great object we have in common.*[28]

## The Flight-Testing Program

The AEA would first go aloft with a kite, the Cygnet I, which was Bell's largest kite ever and made with hundreds of red silk tetrahedral cells on pontoons. One of the team's priorities was to prevent the loss of human life while running their tests, and the key to success in this endeavour was their flight-testing program, which was modelled after Sir George Cayley's 1809 process.

The members of the AEA had united with the perfect blend of skills, personalities and energies to fuel the most productive team of international aviation pioneers to ever challenge the more publicized Wright brothers' efforts. Through their work over the coming years, the AEA placed important bricks into the international wall of aviation knowledge.

# 3 THE FIRST PLANNED FLIGHT

The AEA accomplished many firsts in their eighteen-month existence and made a lasting impact on aviation history. In Bell's words, the association was a "co-operative, scientific association, not for gain but for the love of art and doing what we can to help one another."[29] Their initial goal was to construct a practical airplane that would carry a person and be driven through the air by its own power.

Initial plans were to construct and fly only five machines, with each member personally overseeing at least one project. The first powered aircraft was to be Bell's kite, equipped with one of Curtiss's motors. It was built by December 1907, but Bell wanted to test it as a glider before engine installation, so Selfridge was selected as the first of the team to attempt flight.

There was nothing small about the Cygnet I. It was a tetrahedral kite as big as a house, made of 3,393 winged cells and having a span of 42.5 feet; even if it did fly, Bell's younger colleagues recognized from the start that it would never be practical. Bell would build two more versions of his Cygnet kite, the Cygnet II (flown from February 22 to 24, 1909) and the Cygnet III (flown on March 17, 1912). Although Bell's own powered and manned tetrahedral

*The Cygnet I tetrahedral kite, towed on a barge as Casey Baldwin lies on his stomach in the passenger cell.*

*The Cygnet I tetrahedral kite under tow by the* Blue Hill *steamer with Lieutenant Selfridge in the passenger cell, December 6, 1907.*

aircraft generally did not fly well, he provided essential counsel to and was a vital part of the group as a whole.

Preliminary tests were run by towing the empty Cygnet I behind some boats on the Bras d'Or Lake. On December 3, 1907, the Cygnet I flew briefly as an unmanned kite, experienced some modest damage on landing and required quick repairs for a manned flight trial. Selfridge and Curtiss returned to Baddeck in late November, and Selfridge made the first flight on December 6, 1907, in Bell's Cygnet I. The structure of the kite weighed 47.052 kilograms (104.5 pounds), and even with additional floats for water landings, the entire device only weighed 94.5 kilograms (208 pounds). Towed behind a steamer

*The Cygnet I tetrahedral kite, unmanned trial flights over Baddeck Bay, December 3, 1907.*

*The Cygnet I tetrahedral kite, towed by the* Blue Hill *steamer on Baddeck Bay, December 1907.*

on the Bras d'Or Lake, the Cygnet I, with Selfridge aboard, took flight to an altitude of 168 feet (51 metres).

The Cygnet I, a much larger version of the Frost King, became the first planned manned flight in one of Bell's kites. He chose the name Cygnet from the Latin word, Cygnus, meaning a large, white bird. Mabel wrote of these experiments as being very encouraging. "He knows how to send a kite into the air from the water and how to receive it from the air. [Early stage takeoff and landings] are the two things he wanted solved before he would allow anyone to ride a kite-flying machine."[30]

On December 6, 1907, Mabel stood on the upper deck of the local steamer *Blue Hill*, which the AEA had hired to tow their catamaran barge. On it rested the Cygnet I, looking like a huge wedge of honeycomb with its hundreds of red tetrahedral-shaped silk cells. The kite took to the air with Selfridge on board, immediately soaring high above the water, and it flew with great steadiness in the stiff breeze. Selfridge continued to fly in the Cygnet for over seven minutes, until the large kite started to drift down due to a decreasing breeze that could not be totally offset by the progress of the small steamer. Lieutenant Selfridge lay in a small open area in the middle of the kite (the cockpit) and shifted his weight to control its flight.

When the wind dropped and Selfridge landed on the water, smoke from the steamer obscured the kite line. Unfortunately, because of the blinding smoke that belched from the towing steamer, everyone on the steamer momentarily lost sight of the kite as it landed and no one cut the tow line in time before the floating kite was dragged violently through the water and destroyed, making this the Cygnet I's only flight. When Selfridge felt the fragile craft breaking up, he managed to crawl out of the cockpit and plunge into the icy water. Selfridge came to the surface in the frigid waters of the lake and was quickly rescued and warmed up.

The flight of the Cygnet made Selfridge the fifth passenger on an aircraft in Canada, after Louis Lauriat first flew a balloon in August of 1840 at Saint John, New Brunswick. Lucian McCurdy and Neil McDermid flew second and third on the Frost King in 1905, and Lawrence Lesh was fourth to fly in a glider in Montreal in August of 1907. Bell immediately sent a telegram with the news to the major newspapers and the story was given wide press coverage.

The two kite flights in the Frost King were unmanned because it wasn't known for sure if the tetrahedral kite was capable of lifting a person. The people of Baddeck were so thrilled that the first officially planned flight of a person and kite had taken place in their village, they presented Bell with a silver tray on Saturday, December 7, 1907, to commemorate the event. This experiment ended test flights at Beinn Bhreagh for that year as the Cygnet I underwent repairs, while the group's thoughts shifted to a more conventional aircraft, such as those being tested in Europe at the time.

A double-decker glider would be the next aircraft flown, in Hammondsport, and it was more of a group effort among the four younger men. Under Bell's guidance his group stuck to a step-by-step plan, and the gliding stage needed further trials before advancing to an engine-powered craft. Bell's team knew they had to progress carefully and strive to maintain a high level of safety while transitioning to powered airplanes.

Bell's pilots were some of history's very first true test pilots, as they were truly starting from scratch. Every single survivable flight was a victory, in which the pilot's personal knowledge of the experience was retained for the next potentially dangerous, yet hopefully safer test flight. As basic as it may seem, simply flying higher and doing full turns in the sky was a big step forward for pioneering pilots, who had never before done those things.

# 4 EVOLUTION TO POWERED FLIGHT

When the Cygnet I kite was wrecked in the test flight over Baddeck Bay, it gave the younger men of the AEA an excuse to move winter operations to Curtiss's workshop in Hammondsport, in New York's Finger Lakes region. There they could more easily access the engines required for future AEA airplane designs. Curtiss and the other young members of the AEA were more interested in producing conventional aircraft, like the emerging designs of other aeronautical pioneers around the world, and in pursuing controlled, powered flight. Similar to the space-race to get humans on the moon in the 1960s, this was a period in which the quest was on to produce a reliable passenger-carrying airplane. After the group moved to Hammondsport for the winter, they would design a series of aircraft, a glider and four powered airplanes with more stylish names.

For the next AEA aircraft, Bell's team adopted the basic design of the Chanute-Herring biplane hang glider, which flew well in its 1896 experiments near Chicago and has been copied throughout aviation history. They also used aeronautical data on lift provided by German pioneer Otto Lilienthal, who was known as the "Glider King."

The aerofoils, or wings, had a camber, which is a curvature of the top surface. *Aerofoil* is the term used

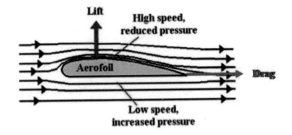

*A cross section of an aircraft wing, detailing airflows and lift.*

to describe the cross-sectional shape of an object that, when moved through air, creates an aerodynamic force. Aerofoils are employed on aircraft as wings to produce lift, a force perpendicular to the air flow that is created as air flows faster over the wing, speeding up to go a longer distance, causing a decrease in pressure.

The wooden upright shafts between the wings were braced by wires. By 1896, Octave Chanute, the American railway engineer turned aviation enthusiast, had concluded that the externally braced biplane offered better prospects for powered flight than the monoplane; however, the weakness of the materials and design techniques available around that time admittedly made it difficult to design wings that were both light and strong enough to fly.

In January 1908, the Bells returned to Washington while the three young men joined Curtiss in Hammondsport; Bell was becoming more of a figurehead and overseer for the AEA. Although Bell only got to Hammondsport at irregular intervals, he kept in close touch with all that was going on by establishing a weekly bulletin, a practice that continued until the association terminated in March 1909. The full set of AEA bulletins contains over two thousand pages of data on aeronautical experiments and can be viewed in the archives of the Bell Museum in Baddeck.

Bell relaxed on the tetrahedral design for the short term and agreed to expand their experimentation to include current biplane designs. In Hammondsport, they constructed a biplane glider as a first step to building aircraft that were not based on tetrahedral kite construction.

During this early period in aviation, the issues of lift, stability and control were not understood, and many flight attempts ended in death or at least serious injury to the pilot. Pioneer designs were often without effective horizontal tails, or the main wings were too small and lacked the curvature that creates lift. A lack of lateral control was a problem, as pilots were still trying to turn aircraft by shifting their body weight from side to side in the gliders. (Some of today's modern hang gliders successfully use this technique to fly safely, which aligns with Bell's notion that a powered kite could be a safe flying machine.)

Bell's team gathered all the information they could on aviation and aeronautical engineering, similar to the literature search the Wrights had done when they began their glider experiments in 1899. In fact, Curtiss and Selfridge wrote to the Wright brothers, asking advice. Usually very secretive, this time the Wrights were as candid with the AEA as the Smithsonian had been with them. They answered questions about engineering and materials, and directed the members to published papers and patents for more in-depth information. They thought well of Bell, and it impressed Wilbur and Orville that he was in charge of the AEA.

Using this information and research, the AEA started experimenting with the French-designed "canard" — a fixed-wing aircraft in which a larger main wing is set back behind a smaller forward horizontal surface, also called a foreplane or stabilizer. The early aircraft designs had no tail, and the small control surfaces in the front reminded the French of a flying duck — hence the name.

In the early 1900s, aviators often stalled their aircraft at low altitudes. When an aircraft stalls, it simply gets too slow and the nose of the aircraft starts to drop toward the ground. The parachute effect of the forward elevator (little wing) in the canard design allowed pilots to make a safer, flatter crash landing instead of a more deadly nosedive toward the ground. These incidents wedded aviators even more strongly to the canard design for a long time.

Like Bell, Octave Chanute was too old to fly himself, so he partnered with younger experimenters. Experiments in 1896 convinced Chanute that the best way to achieve extra lift without a prohibitive increase in weight was to stack several wings one above the other, an idea proposed by the British engineer Francis Wenham in 1866 and realized in flight by Lilienthal in the 1890s. Chanute's contribution to aviation was the introduction of the "strut-wire" braced wing structure that would be used in powered biplanes. He based his "interplane strut" concept on the Pratt Truss, which was familiar to him from his bridge-building career on the railroads.

The biplane design allowed airplane wings to

*An example of the French-designed "canard" type fixed-wing aircraft, which has a larger main wing, set back behind a smaller forward horizontal surface.*

remain rigid and strong while also staying relatively lightweight, as opposed to using the thick metal wing spar that is part of today's aircraft designs. Bell's team often referred to their glider design as the "double-decker."

On January 13, 1908, the members of the AEA began flights with the glider just outside the village of Hammondsport. This preliminary glider had no real tail, although it had an extension that looked like an early attempt at constructing an elevator. It lacked a tail rudder, although smaller vertical fins can be seen on the outer wing tips, and there were small landing skids on the bottom wing.

They completed approximately fifty unpowered gliding flights over two months on the snowy hills surrounding Hammondsport. The initial flights measured only a few feet in length as the pilots struggled to find the balance point in the fledgling weight-shift craft. The glider was crude, so the soft snow provided

*Pilot Baldwin (standing in the center) at the edge of a rise, preparing to launch the glider for a test flight on the south side of Pleasant Valley, Hammondsport, New York, in February 1908.*

welcome protection from the impact on landing.

The AEA designed a very light biplane structure that fitted over the shoulders yet allowed the experimenter's feet to remain free for running. The would-be birdman raced down a steep slope, hoping for sufficient speed to lift him off the ground, and occasionally he made it into the air, but the glider still lacked stability. Adding a proper tail structure improved matters considerably. After this, flights of 90 to 120 feet were not uncommon, and on one occasion McCurdy recorded a flight of over one hundred yards. Extensive trials of the glider provided not only valuable performance data, which was folded back into the evolving designs, but it helped the AEA pilots develop essential piloting skills. By mid-March 1908, they concluded work with the glider as their flights were reaching the hundred-yard mark. These glider flights were mostly carried out by Baldwin and McCurdy, who stated that this experience served a very useful purpose by giving them a feel of the air,

*Glider testing in February 1908 by the AEA from the hills on the south side of Pleasant Valley, Hammondsport, New York.*

and it taught Bell's team how to balance the craft when it was airborne.

Prior to the glider flight test period, the members of the AEA had already determined that they would have to design a succession of machines to keep up with the damage inflicted during testing. They expected they would be learning as they went, until they would be able to master all of the factors necessary to build, power and manage the flight of an airplane. Early in January, as they finished work on the glider, they started on the design of their initial airplane, a modified glider that had the addition of a Curtiss motorcycle engine and a propeller.

German glider pilot Lilienthal once said, "To invent an airplane is nothing. To build one is something. But to fly is everything."[31] There were four major challenges to getting people in the air up to this point. The first two were finding the money and the time to do such experiments. The third challenge was obtaining an engine powerful, light and dependable enough to keep the airplane aloft for any length of time. However, the AEA had these items at their disposal now. The last and most critical challenge was teaching oneself when flying was in its infancy — every single takeoff and landing without a loss of life was a significant accomplishment.

## Red Wing (Aeroplane No. 1)

Bell's associates were able to innovate, as each member was responsible for developing a project while the rest of the team helped out. The successful aspects of each project were retained for the next endeavour and eventually incorporated in the development of future flying machines. The difficulties of propulsion and determining the most efficient wing shape had already been solved by other pioneers, for the most part. As the association's leader, Bell asserted that Selfridge, having risked life and limb piloting the Cygnet I, had earned the right to the next aircraft design, so Aeroplane No. 1 would be Selfridge's creation. It was dubbed the "Red Wing" after the bright red silk used to cover its wings, as Bell had long since determined that the colour red shows up better in black and white photos. The Red Wing was finished in three months at Curtiss's shop in Hammondsport.

The canard designed aircraft contrasts with today's conventional aircraft, which have a small horizontal surface or tailplane behind the main wing. Pioneering aviators favoured a canard format because they thought that German pilot Lilienthal's tragic accident — in a glider with an aft tail — had been due to Lilienthal having a lack of control. Lilienthal died after he stalled. In a canard design, the pilot would be able to see the control surface; as a result the AEA used this design on the Red Wing aircraft.

Selfridge's Red Wing was a double-decker biplane with a moveable elevator ahead of the main wings

*A front quarter-view of the Red Wing airplane being inspected by workers prior to the first flight.*

*Wreckage of the Red Wing.*

and a fixed stabilizer behind the main wing. A moveable tail rudder was provided, but no means of lateral control had yet been introduced to the AEA models. Like the gliders, the Red Wing was still dependent on the pilot shifting their weight to steer the craft. Being flown from the ice, the Red Wing used a skid runner undercarriage for takeoff, powered by a forty horsepower Curtiss motorcycle engine with a push-type propeller. It had a wingspan of forty-three feet, four inches, and its length was twenty-six feet. The fabric wings stretched over spruce and steel ribs and were cross-braced by wires in the Chanute Pratt Truss style. The Red Wing was ready for a trial run on March 10, 1908; however, Selfridge and Bell were on business in Washington, so the honours to test-fly the creation were handed to Baldwin.

On March 12, 1908, Baldwin piloted the aircraft off frozen Keuka Lake near Hammondsport. It was the first public demonstration of a powered aircraft flight in the US, as well as the very first powered flight by a Canadian pilot anywhere in the world. A contemporary account described the Red Wing flight as the "First Public Trip of Heavier-than-air Car in America."[32] Reports entitled "Views of an Expert" stated that Bell's new machine was "shown to be practicable by flight over Keuka Lake, Hammondsport, New York, March 12, 1908."[33]

The aircraft ran across the ice on its steel skids and did an "unassisted" takeoff (no rail or catapult) to cover 319 feet (97 metres) at a height of around 20 feet (6 metres) in 20 seconds, before it crashed as a portion of the tail gave way. The second flight, on March 18, also ended in a crash, but it covered only 121 feet (37 metres). Both tests proved the need for lateral control. The Red Wing was damaged beyond repair, and Baldwin claimed when the tail broke off he did not know it had broken at all, causing him to feel that they should have moveable parts in front of the pilot where the pilot can see them. For a while longer at least, this became another argument for keeping the canard design.

After the crash of the Red Wing, Baldwin and Bell realized the one factor that prevented them from achieving a more successful flight: the pilot's lack of ability to balance and control the craft once it was airborne. It was obvious to them that using the weight-shifting method to fly wasn't working. They also recognized that they had expected the rudder to behave like a boat rudder does in two dimensions. In 1908, most aeronautical investigators still regarded flight as if it were not so different from surface locomotion. They thought in terms of

*The White Wing leaves the ground on its first flight on May 18, 1908, as people stand watching in the foreground.*

a ship's rudder for steering, while the flying machine remained essentially level in the air. The idea of deliberately leaning, or rolling, to one side seemed either undesirable or did not enter their thinking, and this lack of understanding on the theory of flight led to numerous crashes, including the crash of the AEA's Red Wing. It was gradually realized that an aircraft flies in three dimensions, as opposed to a boat, which turns while staying relatively flat on a two-dimensional plane. Bell's most significant contribution to the development of aviation was his ability to analyze aviation failures and to propose remedies, as demonstrated by his suggestion of using ailerons following the Red Wing's crash.

An aileron (French for little wing) is a hinged flight control surface, forming part of the trailing edge of each wing of a fixed-wing aircraft, and they are used in pairs to control the aircraft in roll. Between 1868 and 1909, a number of pioneers experimented with methods to create lateral balance and the ability to safely turn airplanes. It is entirely possible that some of these people came up with similar ideas at the same time and were unaware of each other's developments in other parts of the world.

Once, while swimming the backstroke in the Bras d'Or Lake, Bell realized he was balancing himself with his arms extended like aircraft wings while using his hands as ailerons in the water. Water and air follow many of the same principles, with one being denser than the other, and knowing this led Bell to use ailerons to balance the wings of an aircraft.

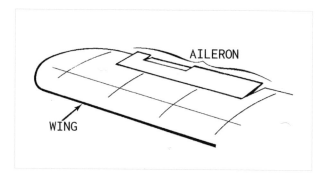

*An aileron is a hinged flight control surface, forming part of the trailing edge of the wing. Ailerons assist in turning the aircraft.*

Bell had apparently been subscribing to the French journal *L'Aerophile*, which was published in the late nineteenth century. Assuming that he received those copies during their years of publication, it is more likely the way that the AEA came up with the idea to add ailerons to their airplanes. In an aileron article in *L'Aérophile*, they published illustrations of ailerons on Esnault-Peltérie's glider in June 1905, and its ailerons were widely copied afterward.[34]

In a letter to Baldwin, Bell recommended the idea of "moveable wing tips" and he suggested they be controlled by cross wires worked by the pilot. Bell and the AEA received a US patent for the device, "No. 1,011,106 Flying Machine — Bell, Baldwin, McCurdy, Curtiss, Selfridge," on December 5, 1911.

The history on lateral control for an aircraft (the ability to remain level or turn), goes back to 1868, when a British inventor, Matthew Piers Watt Boulton, first patented the aileron. However, for many years the patent was relatively unknown. Another design used to turn aircraft, which looks and operates a lot like ailerons, is called "wing warping," where the wing tips actually warp like a bird's wing tips flex. Wing warping was jointly owned and patented by Louis Mouillard and Octave Chanute in 1897. Regardless of the 1868 Boulton patent, the Wright brothers' Ohio patent attorney, Henry Toulmin, had filed an expansive wing warping patent application, and on May 22, 1906, the Wrights were granted US Patent 821393.[35] Bell's patent in 1911 for the aileron was cancelled by US courts because it was said to be an infringement on the Wrights' wing warping — yet Mouillard and Chanute had already patented wing warping. It's interesting that the Wright brothers received and kept a patent on wing warping for as long as they had it, considering Mouillard and Chanute patented wing warping first and the aileron was patented twenty-nine years before wing warping. The Wright's patent would have been an infringement on both the aileron patent in 1868 and the wing warping patent in 1897, if faster modes of communications had existed in 1906.

## The White Wing (Aeroplane No. 2)

After analyzing the flight of the Red Wing, the group found many ways to refine their second aircraft, which was Baldwin's White Wing. Like the Red Wing, it was a biplane with a front elevator, rear stabilizers, a rudder connected by wires to a three-axis steering wheel and a rudimentary aileron system controlled by the body motions of the pilot. The White Wing's most important innovation, the addition of moveable lateral control surfaces on all four wing tips, qualified as North America's first ailerons and the AEA's solution to lateral stability. Bell's team felt that wing warping was a less efficient application and that the AEA "wing tips" were an important step in extending the airplane's practicality.

With the White Wing's ailerons, Curtiss claimed mechanical and control differences existed in this aircraft design because the method of control reflected his motorcycle experience — a yoke embraced the pilot's shoulders — and a cable was used to engage the ailerons. The pilot would lean into a turn or away from an undesired dip, imitating the movements of a motorcycle operator, resulting in the proper control movements being automatically applied without the pilot having to think about it. The movements were instinctive, and the results were elegant, a simple transition for anyone who could ride a bicycle.

While other aircraft at the time almost universally had two wheels or skids, the White Wing showcased a three-wheel, rubber-tired undercarriage, another carry over idea from Glenn Curtiss's motorcycle sidecars. It was an unusual aircraft in its day, being the first airplane in America to use a tricycle landing gear, but it made unassisted takeoffs easier (no use of rails or catapults). As late as 1908, wheeled undercarriages on aircraft were still considered unusual. Aeroplane No. 2 employed the engine from its predecessor, the Red Wing. Baldwin flew the White Wing first on May 18, 1908; he was followed by Selfridge, who made the first powered flight by someone from the US military.

Curtiss, the motorcycle man, celebrated his thirtieth birthday by making his first flight on May 21, steering both right and left, with the airplane in perfect control at all times. After many reassessments and revisions, the plane was ready for flight on May 22, 1908, and with Curtiss at the controls again, it flew successfully at a height of about ten feet for nineteen seconds, and covered a distance of 1,017 feet (310 metres). On May 23, however, McCurdy wrecked the White Wing beyond repair after flying 549 feet (167 metres). He sustained a knee injury in the crash, but nothing else was lost, as the association had learned a lot already from the White Wing's four flights. Despite its short existence, the aircraft introduced important advancements with wheels and ailerons that still remain on modern aircraft.

The advancements made by the AEA with the White Wing set the stage for Bell's team to start publicly overtaking the performances of more well-known aviators. Bell's team's ideas would eventually be adopted by other North American aircraft manufacturers as the preferred way to build aircraft. After the White Wing was damaged, the AEA turned their full attention toward building Aeroplane No. 3.

# 5 BELL'S TEAM SURPASSES THE WRIGHTS

The AEA was rapidly gaining ground on the Wrights after they started using ailerons to enhance lateral stability. Aeroplane No. 3, the June Bug, included a new feature at the front of the craft: a moveable canard design that allowed the pilot to control the elevation by making subtle up and down adjustments to the front elevator, and improved the handling of the airplane in flight. This feature coupled with the ailerons made the June Bug the most stable of all AEA aircraft thus far.

Since they were learning everything from scratch, the first attempts in May failed to get airborne until Bell realized that the fabric wings were too porous.

Thus, Curtiss's June Bug was the first airplane to have its wings doped with a surface covering that would provide more lift to the aircraft. Aircraft dope is a plasticised lacquer applied to fabric-covered aircraft that tightens and stiffens the fabric stretched over the airframe, rendering it airtight and weatherproof. Doping the June Bug's wings solved the problem, and a yellow ochre was added to the shellac and varnish combination that was painted on the porous white cloth wings to give the aircraft a yellow hue.

To the casual eye, the AEA's Aeroplane No. 3 appeared to be simply a modification of the White Wing with the wing surface reduced and the aileron

*A cloud of dust is raised as the June Bug takes to the air on its first flight on June 21, 1908.*

surface increased. The frame and wheelbase were stretched a little, and the cloth windscreen was removed — lead designer Curtiss liked to see where he was going. By mid-June, the June Bug was under a tent at Stony Brook Farm in Pleasant Valley, just south of Hammondsport, where vintner Harry Champlin had loaned his horse track for a make-do runway. In spite of looking much like the White Wing, minor improvements dramatically increased the June Bug's overall performance. Its maiden flight was taken on June 21, 1908, and covered 456 feet (139 metres) in eleven seconds. By June 27, on its seventh flight and with Curtiss at the controls, the June Bug flew for sixty seconds and covered a distance of 3,420 feet (1,042 metres). Curtiss flew the June Bug for the first nine flights between June 21 and 29, 1908. Flights were now limited only by the length of cleared fields, because early pilots were still being cautious by not flying too high off the ground. Pilots were still flying low level hops as Curtiss surpassed a kilometre.

Progress encouraged the AEA members to compete for the *Scientific American* trophy. They entered the June Bug in a competition to take place on July 4, sponsored by *Scientific American*, for the first public flight over a kilometre-long course.

Selfridge announced, "We have telegraphed and telephoned Secretary Aero Club of America that we are ready to try for the *Scientific American* cup. Hurrah!"[36] This announcement caused a stir in New

*The Wright brothers' first runways were the monorail type.*

York City, where the Aero Club and *Scientific American* both had headquarters. The magazine had offered a trophy and prize money aimed at the Wrights for an unassisted takeoff followed by a one-kilometre straight flight — which Curtiss had just beaten in an unofficial effort. The entire competition had been the brainchild of the magazine's publisher, Charles A. Munn, who felt bad about how his magazine had treated the early reports of the Wright brothers, and who was making a virtual gift of the prize and the money (a $2,500 cash award) for the first public flight of one kilometre to the Wrights. All they had to do was complete the flight to claim it.

In secrecy, the Wright brothers were rumoured to have flown twenty-four miles in a circular pattern, but their efforts were not public and remained unofficial; moreover, they were still using a monorail launcher, which was a track of 2x4s stood on their narrow edge. Officials offered to hold off awarding Curtiss if the Wrights would quickly make an effort. Orville responded that they could substitute wheels for skids, but he expressed his confidence that future airplanes would continue to be launched from special apparatuses rather than from wheels on the ground. Their reluctance to give up the launching rail was likely connected to their fears that changes might affect their patent protection. Finally, their preparations for US Army trials also apparently made the trophy effort a great inconvenience for the Wrights.

In the end, the Wrights refused (even declining the written pleas of Munn), claiming that their plane did not meet the qualification of taking off unassisted. The truth was that the Wrights would probably have turned Munn down anyway, and this left the field open for the AEA. The Aero Club and *Scientific American* resignedly returned to Curtiss and McCurdy, who further exasperated them by insisting that the trial take place in Hammondsport rather than in New York City. Curtiss, usually reserved in public, jubilantly told his townsfolk, "We'll fly the June Bug on the Fourth of July. Advertise it, invite everybody interested in flight. Draw a crowd to Hammondsport and we will prove to the world that we can really fly."37

On July 4, 1908, the entourage of officials from the Aero Club, *Scientific American* and the press assembled around a farmer's field just west of the village of Hammondsport. With photographers, officials and a huge crowd of spectators in place, the June Bug's engine roared to life as Curtiss taxied the airplane onto the Stoney Brook Farm's dirt horse track runway. Curtiss gunned the engine and the airplane lifted as officials carefully marked the point where the wheels left the ground. Flying along at an altitude of some twenty to thirty feet, Curtiss kept the June Bug aloft for a distance of 6,000 feet (1,829 metres) in a flight that lasted 102.5 seconds. Weaving a little to avoid obstructions, he passed the red one-kilometre flag and kept going. He flew the machine as far as the field would permit, regardless of fences or ditches, and it was only as he was running out of space for straight line flying that Curtiss brought the June Bug carefully down in a corn field just short of a copse of trees. The crowd went wild as the plane landed well past the measured one kilometre distance required to win the *Scientific American* trophy and claimed the prize, much to the embarrassment of Munn. Bell's team was overtaking the Wright brothers in full view of the general public. Meanwhile, in the nearby township of Pleasant Valley, they threw the wine cellars open to celebrate.

The younger of Bell's daughters, Daisy Fairchild, described an electrified atmosphere when she said,

> *In spite of all I have read and heard, and all the photographs I have seen, the actual sight of a man flying through the air was thrilling to a degree that I can't express. We all lost our heads and David shouted and I cried and everybody cheered and clapped and engines tooted . . . The banks were crowded with spectators . . . The machine rose beautifully and flew by us . . . All sorts of pictures were taken to [sic] and the air was full of the click click of shutters — there were moving picture cameras and Kodaks of all sizes . . . The first flight had raised excitement to the boiling point . . . I don't think any of us knew quite what we were doing. One lady was so absorbed as not to hear the coming train and was struck by the engine and had two ribs broken . . . the people were so excited and yelling, you could see their mouths going, although it*

*The AEA's June Bug wins the* Scientific American *trophy on July 4, 1908, in Hammondsport, New York.*

was difficult to hear them due to the engine noise.³⁸

David Fairchild, Daisy's husband, sensed the historical and future implications of the day when he wrote of the June Bug event in his book *The World Was My Garden*, and said,

> *That brief afternoon at Hammondsport had changed my vision of the world as it was to be. There was no longer the shadow of doubt in my mind that the sky would be full of aeroplanes, and that the time would come when people would travel through the air faster and more safely than they did on the surface of the earth.*³⁹

Bell was already burning up the wires to his patent lawyers. It turned out to be a sensible move because Wilbur Wright wrote to Orville from Europe five days later and stated his opinion that their own Flyer aircraft could not be made practical without using features of the AEA's patent, and he recommended that they ask Curtiss whether he would like to take a licence to operate under the Wright patent for exhibition purposes. The Wrights would not offer any manufacturing rights. The curtain was going up on a decade of acrimonious patent wars, and the brothers were shrewdly focusing their attention on Curtiss, rather than on the highly revered Bell.

Although Bell could not be present at the June Bug's flight, he telegraphed exuberant congratulations: "Hurrah for Curtiss! Hurrah for the June Bug! Hurrah for the Aerial Association!"[40]

The Wrights informed Curtiss that they would sue him and the AEA. After extensive experience with telephone patent court battles, Bell was doubtful about what the outcome of a patent fight with the Wrights would bring and he wanted no part of being on the losing side. The fact is, he did have his lawyers inspect the June Bug for possible patents and received a discouraging report. Pioneering aviator Lawrence Hargrave wrote in 1893: "Workers must root out the idea [that] by keeping the results of their labors to themselves[,] a fortune will be assured to them. Patent fees are much wasted money."[41]

The AEA was not expecting to become rich. Still, the success of the AEA that spring and summer had cast some serious doubt as to whether the Wrights were the best airplane manufacturers available. By the end of 1908, after Bell's team had completed over 150 flights in the White Wing and June Bug without mishap, the Wrights recognized them as serious competitors.

The experimental phase of the AEA was over with the success of the June Bug flights. They had gone beyond the "minutes" stage of flying, and they were now contemplating heading into the "hours" stage. In order to do this, they had to graduate from the quick-heating, air-cooled engine to one that was water cooled. As temperatures were quickly dropping in the fall of the year, they also realized that some kind of anti-freeze fluid would soon be necessary to prevent the engine block from being damaged. Curtiss immediately started production of an eight-cylinder, water-cooled motor for Aeroplane No. 4.

The time spent waiting for the completion of this new engine gave Bell's team time to think about a whole new approach to takeoffs and landings. Bell, ever safety conscious, believed takeoffs and landings were the riskiest part of flying and decided that practising them on water would be less dangerous than on land.

# 6 FLYING ON WATER

Trials with the Red Wing, White Wing and June Bug aircraft had demonstrated the need for more power, and while Curtiss was working on a water-cooled motor for the Silver Dart, Baldwin and Bell took the time to experiment with hydroplaning pontoons.

Bell was detained in Washington in the spring of 1908, resulting in delays to the original plan to resume work on the tetrahedral airplane, the Cygnet II, at Beinn Bhreagh. Their work now focused on hydroplanes, to increase the efficiency of takeoff of an airplane from water. The notion of a water airplane cropped up spontaneously within the group; after getting planes to fly off the ice and land, it was only natural for the AEA to try landing planes on water. After July 4, 1908, Bell divided the association members to carry out work at Beinn Bhreagh and Hammondsport simultaneously. Curtiss and McCurdy remained in Hammondsport to construct Aeroplane No. 4 while Baldwin proceeded to Beinn Bhreagh to give Bell assistance with his tetrahedral structures. Lieutenant Selfridge was ordered back to Washington by the US Army.

Bell and his team recognized that the airspace over open water provided large natural areas that were free of obstacles such as trees, hills and buildings. Like flying over the ice, open water allowed pilots to

*A rear view of the Dhonnas Beag hydrofoil, October 28, 1908, in Baddeck.*

continue to fly at lower altitudes where they felt safer and could fine-tune their experimental airplanes to become more reliable. There was a major difference between flying over land and water — if one were to crash, splashing down was usually much more survivable than hitting the hard, unforgiving ground, assuming the pilot could swim.

Bell and Baldwin thought that using hydrofoils would help their floats attain higher speeds, an essential requirement for flight. "There would be three phases of operation as speed increases during take-off: first, operation on the displacement hull; second, running on the hydroplanes; and third, flight on the surfaces of the cell bank [the wing area of the tetrahedral] . . . described as . . . this aerohydric trinity of a boat, a hydroplane and an aeroplane."[42]

*Baldwin, Bell and an unidentified worker test-ride a new hydrofoil on Baddeck Bay.*

Hydrofoils are similar in appearance and purpose to aerofoils (or wings) used by airplanes. A hydrofoil is a lifting surface, or foil that operates in water, consisting of a mini-wing (finlike) structure mounted on struts below the hull of a watercraft. As a hydrofoil-equipped watercraft increases in speed, the hydrofoil elements below the hull develop enough lift to raise the hull out of the water, which greatly reduces hull friction, or fluid resistance. This provides a corresponding increase in speed without the need for a much larger engine. Watercraft that use hydrofoil technology are commonly referred to as hydroplanes.[43]

Bell and Baldwin visited with Italian inventor Enrico Forlanini and rode in his experimental hydrofoil boat on Lake Maggiore in Italy. When they returned to Baddeck, Bell employed about forty people to help build a number of preliminary versions of these experimental air and water craft. As Baldwin studied the work of Forlanini and began testing models, his and Bell's work evolved toward hydrofoil watercraft or flying boats. Bell thought that the boat should form the body (fuselage) of the airplane. Starting its takeoff from the water, Bell hoped the flying boat could develop enough speed under its own motive power to rise from the water into the air. However, remaining very safety minded, Bell felt that there was no need for the flying boat to rise to any great elevation above the water. Planning on flying at an elevation of only a few feet above the surface of the water, Bell believed the danger in the event of an accident would be reduced, for it would drop into the water instead of on land. Bell thought that, should their Aeroplane No. 6 succeed in rising from the water under its own motive power, it would undoubtedly become an epoch-making machine.

In August 1908, under Baldwin's direction, a special hydroplane was produced and christened the Dhonnas Beag (Gaelic for Little Devil); it would eventually become the first Bell-Baldwin self-propelled hydrofoil. Through reducing its weight by omitting the engine and passenger, Baldwin succeeded in lifting the Dhonnas Beag completely out of the water on its hydroplanes by towing it. The body formed a boat with out-riggers so that it could float upon the water and rise like a water bird. After Baldwin placed hydroplanes under the boat body to assist in this process, they planned for a tetrahedral kite structure to constitute the aerial part of what would become Hydro-Aeroplane No. 6.

## Water Trials Continued

The term "Skidoo," often associated with snowmobiles, is actually a term that Bell applied to a motorboat used to tow hydroplanes and

*Workers on a dock in Baddeck preparing to test the Dhonnas Beag hydrofoil.*

*A close-up quarter view, looking down, at the Cygnet II tetrahedral kite on the sheltered waters of Baddeck Bay, near Beinn Bhreagh.*

*In the fall of 1908, a couple of Oionus-type tetrahedral kites, like the one seen here, were test flown over the Bras d'Or Lakes by Bell.*

hydro-aeroplanes on Baddeck Bay in 1908. They planned to convert the hydroplane into a hydro-aeroplane after succeeding in lifting the boat-like pontoons successfully out of the water. Twice on October 8, 1908, the Dhonnas Beag, while being towed and carrying two passengers each time, lifted its pontoons out of the water and ran while hydroplaning. Bell wrote in the AEA bulletin, "On October 29, 1908, the Dhonnas Beag, carrying Mr. Baldwin and the Curtiss No. 2 engine, rose completely out of the water on her hydroplanes when propelled by her own power instead of being towed by the Skidoo."[45] In recognizing the ways that this experiment advanced the development of Aeroplane No. 6, Bell said, "This marks a decided and vigorous step forward . . . the foundation for radical change and wonderful advance in things nautical as well as aeronautical."[46]

On November 6, 1908, they attempted to fly the White Oionus kite, which was to form the aerial structure of Hydro-Aeroplane No. 6, on top of the Dhonnas Beag's base. Unfortunately, Bell and Baldwin decided to fly the White Oionus first without a tail; this caused the machine to jump around so much in the air that the line snapped and the kite smashed as it drifted away with the wind. The damage was considerable and it was decided that it would be better to construct another kite. Thus, even though the Dhonnas Beag rose out of the water on its own power, it cannot be considered a hydro-aeroplane, as the combination of the White Oionus and Dhonnas Beag, which was to be Aeroplane No. 6, was never able to be constructed.

Realizing that bigger wasn't necessarily better when it came to testing the airborne tetrahedral kites, a half-sized model of Aeroplane No. 5 (also known as the Cygnet II) was constructed, towed and flown as a kite over water on October 12, and again on November 27, 1908. The experiments went with

*A modern stepped-bottom pontoon design facilitates takeoff by reducing suction friction, because at high speeds the rear of the pontoons are out of the water first.*

the precision of clockwork, and the kite landed without incident. Testing of Aeroplane No. 5 would not resume until February 22 to 24, 1909 on the ice of Baddeck Bay.

A kite, simply called the White Kite (which was different from the White Oionus), crashed and smashed during testing on December 8, 1908. Another Oionus-type kite, known as the White Victor, was flown the same day when Bell was unwilling to lose the opportunity of employing a good kite breeze. This also resulted in damage when it made its last flight on a hundred-metre line.

Meanwhile, back in Hammondsport, from October to November the June Bug was modified by adding "catamaran-like" floats to its underside to create a seaplane. Renamed the Loon by McCurdy, it was put on twenty foot, twin, wood-framed pontoons made from California redwood and covered with rubber oilcloth. Attempts to fly it began on Keuka Lake on November 28. The modified airframe of the June Bug, fitted with another Curtiss engine, became the first hydro-aeroplane tested in the US. Although the aircraft could achieve speeds of up to twenty-nine miles per hour (forty-seven kilometres per hour) on the water, it was unsuccessful in its takeoffs on November 28 and 29, 1908. After a 1,200 foot (366 metre) run without the pontoons rising from the water, McCurdy noted that the suction of the flat-bottomed floats was responsible for holding the Loon down.

Modern aircraft float designs include a stepped bottom as opposed to a flat bottom to facilitate takeoff. The airplane is given full takeoff power until the plane gets "on the step" — when the hull is supported on the water's surface as opposed to ploughing through it, similar to a person water skiing. As speed and lift increase, the seaplane lifts onto its step, barely skimming the water, with friction at a minimum.

On January 2, 1909, the Loon unexpectedly sank in twelve feet of water after one pontoon developed an undetected leak. McCurdy and Curtiss displayed a sense of humour about the event when they telegraphed Bell, telling him that the submarine test was successful. It would not be until June 1, 1910, that successful water test flights would resume for Bell's team, then using the Baddeck No. 2 Aeroplane.

Years after the Baddeck No. 2 was tested on floats, Bell encouraged Baldwin to continue to design and build more flying boats. Baldwin built a hydrofoil boat at Beinn Bhreagh and designated it the HD-1. Its aerial propeller and short biplane wings for extra lift made it look like a stunted seaplane skittering across Baddeck Bay in a series of unconsummated takeoffs. After being redesigned and rebuilt, it was tested again before it cracked up from unknown causes, but not before reaching a top speed of 50 miles (80 kilometres) per hour. A series of HD models were built in Baddeck — the HD-2, HD-3 and HD-4. Few people realize Bell was also the force behind the world's fastest boat, the HD-4, a futuristic-looking "winged watercraft" that set a 1919 speed record of 70.86 miles per hour, which stood for more than a decade. Curtiss would eventually be the one to build flying-boat aircraft in great numbers.

# 7 WHAT HAPPENED IN BADDECK?

Living his senior years in Baddeck with financial freedom, Bell could work on the aviation issues he was so keenly interested in. The first four motorized AEA airplanes were all built, tested and flown in the US. Each airplane taught them something new about flight, and Bell's team continued to incorporate all of their insights into each successive craft. The fourth airplane in the series, the Silver Dart, was completed on October 31, 1908, and first flew on Sunday, December 6, 1908, in Hammondsport, making four flights that first day. Bell felt strongly that the new Silver Dart should be test flown in Canada, too, and the AEA agreed.

It was shipped via rail from Hammondsport through Buffalo, New York, to Niagara Falls. Canada Customs initially prevented the Silver Dart from entering the country, but later allowed entry after intervention from the mayor of Baddeck and the premier of Nova Scotia led Canadian authorities to reduce the import tariff on the aircraft. Entry to Canada was permitted on the condition that the aircraft not remain in the county more than two years.

Following the success of the AEA in Hammondsport, Bell felt he still had unfinished business with his original kite-based aircraft. When the Silver Dart first arrived in Baddeck it was

*The Cygnet II tetrahedral airplane on the ice of Baddeck Bay with McCurdy at the controls, February 22, 1909.*

reassembled in a temporary shelter on the shores of Baddeck Bay, minus the engine. Following Bell's instructions, the mechanics did not immediately install the Curtiss engine in the Silver Dart. Instead, the engine was mounted in Bell's Cygnet II kite, an airplane-like device with a canard elevator system at the front and tetrahedral wings.

## AEA Cygnet II

Aeroplane No. 5, the Cygnet II, was an extremely unorthodox early Canadian aircraft, with a wall-like "wing" made up of tetrahedral cells. The Cygnet II had several Silver Dart components such as a motor mounted at the rear, and was a newer, powered version of the Cygnet I tetrahedral kite designed by Bell in 1907. The Cygnet II looked like a smaller copy of the original design, except for a skid undercarriage and a Curtiss V-8 engine. It had a wingspan of only twenty-six feet, four inches, compared to the Cygnet I's forty-foot wingspan.

All attempts to fly the Cygnet II with the Silver Dart engine on February 22, 1909, met with failure. Unfortunately, during later testing in December 1909, the Cygnet II, which weighed 950 pounds, found itself ploughing through slush and was not able to generate enough power to lift itself off the ice.

On February 22, 1909, the Curtiss engine was remounted in the Silver Dart in readiness for a flight attempt the following day. After the successful tests in Hammondsport, the frail-looking craft was placed on the ice of Baddeck Bay.

The high-speed ice-level tests conducted on Baddeck Bay with the Cygnet II airplane on February 22 caught the attention of the town's population, and by the early morning hours of February 23, news had spread throughout the town that local boy Douglas McCurdy was going to take a more successful flying machine into the air over Baddeck Bay at approximately 3 p.m. There was such a buzz about the impending flight that a school holiday

*Men on skates wheel the Silver Dart into takeoff position on the ice of Baddeck Bay.*

was declared for the balance of the day. The flight was a great public event, bringing people in from all directions, and Curtiss and his wife arrived from New York. Families started to gather as early as noon to claim a spot on the frozen Baddeck Bay, and by 1 p.m. the cold weather had prompted crowds to huddle in small groups while children skated around in fervent curiosity. Many businesses closed and several hundred Baddeck villagers gathered on the ice. The town of Baddeck was nearly empty as onlookers, correspondents and invited guests in great numbers waited to witness the experiment. With only very light winds at ground level, there were perfect conditions for a test flight in a light aircraft.

Due to Bell's desire to have the Silver Dart flown in Canada, the aircraft took its rightful place in the aviation history books. Bell had planned this flight to qualify as the first powered, heavier-than-air machine to fly in the Dominion of Canada and the entire British Empire — which, in 1909 stretched to over 192 nations, covered one third of the world's land mass and had a population of 458 million people.

On October 16, 1908, Samuel Franklin Cody made a flight of 1,390 feet (424 metres) over Laffin's Plain in Farnborough, England, which is recognized by the Royal Aero Club as the first powered flight in England — but not in the British Empire, because the British Empire consisted of territories other than the United Kingdom. Bell knew this, and as a result, Baddeck's residents would be the first to witness this historic flight.

Take a mental step back in time and imagine that today is February 23, 1909, and you have gathered on the ice of Baddeck Bay with hundreds of excited Baddeck residents. You're in a quaint Cape Breton town, where it was very unlikely that major historic events would ever occur, yet this is the part of the world where Bell preferred to live. Bell's contribution to the modern world and its technologies are enormous, and he had become sort of a village godfather in Baddeck, which, at the end of the nineteenth century, was a remarkable community, full of Scottish enterprise and learning.

At 3 p.m., the Silver Dart was rolled toward the outer bay and positioned facing into the wind a mile from the laboratory at Beinn Bhreagh. Mr. and Mrs. Bell sat bundled under a blanket in a horse-drawn sleigh along with the assembled crowd, anxiously awaiting the flight attempt. McCurdy was ready to fly but Bell remained cautious as he recalled witnessing the crashes of Langley's aircraft in Washington. Bell didn't want the flight to proceed until Dr. Dan MacDonald had arrived. Once the doctor was present, McCurdy climbed onto the Silver Dart and sat as cheering crowds surrounded the biplane. As the crowd began to congregate in front of the Silver Dart and along its proposed line of advancement, it became obvious that it would be necessary to appoint police in order to keep the natural ice runway clear for takeoff. Not having seen an airplane before, the locals didn't know it needed a long run to fly.

The Silver Dart took one false start — a gasoline line broke just before liftoff, after the craft had travelled 100 feet (30 metres). This caused a two-hour repair delay, until 5 p.m., which meant the sun was setting as the plane took off on its second attempt. As the engine was about to be started, the light wind quickly shifted from the southeast and started to blow from the northeast. This prompted a relocation of the aircraft by laboratory staff farther up the bay for a takeoff in the direction of St. Ann's Harbour.

Seeing the Silver Dart fly in Baddeck excited everyone, including Mabel, as she described the flight to her daughter Daisy:

> *Another perfect day. The Silver Dart made a short flight, coming down because the land was near, she had to go across the bay on account of a baby wind. We all pleaded hard with Daddysan* [nickname for Mr. Bell] *for another flight but he was firm. It was the first flight of an airship in Canada and he would take no chance of disaster to spoil this first success. All Baddeck was out in sleighs or on skates. We had to wait two hours and they whiled away the time with horse races up and down the smooth hard ice. There must have been fully thirty sleighs within a small compass on the ice. Can you realize people dashing up and down feeling perfectly secure with only a foot or so of ice between them and about forty feet of water. I do just love Baddeck these glorious winter days —you poor Southerners know nothing of their exhilaration.*[47]

John McDermid's one-horse-sleigh towed the aircraft along the ice, assisted by about eight of Bell's laboratory staff, most of them wearing ice skates, who helped by pushing the wings and tail. The Silver Dart was moved to the outer bay and placed in a position about one mile from the Beinn Bhreagh shore facing the northeast wind where it made a start at a spot just off Fraser's Pond.[48] Baddeck Bay runs in a northeast direction, providing plenty of takeoff space for the light Silver Dart to get airborne, as the dense cold winter air provided excellent lift for the aircraft's wings. McCurdy gained valuable hands-on experience in the Silver Dart, flying three of the fourteen reported short test flights at Hammondsport. Completing his first Canadian flight on February 23, 1909, in Baddeck he said,

> *The whole scene is still very vivid to me. It was a brilliant day in more ways than one. The sun was glaring down on the ice of Lake Bras d'Or, which is near Baddeck. The town had turned out in a festive mood, done up in ear muffs and heavy fur hats. The town, by the way, consisted largely of very doubt-*

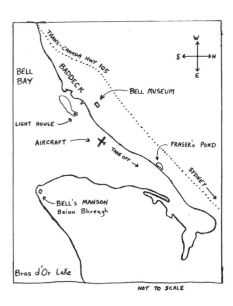

*Fraser's Pond, the starting line for the Silver Dart's takeoff.*

ful Scotsmen. Most of them were mounted on skates — the kind you strap to your feet. They didn't say much, they just came to wait and see. The aircraft . . . was surrounded by people. During the early afternoon it was wheeled into place. The propeller was cranked, and with a cough, the motor snorted into place. I climbed into the pilot's seat. With an extra special snort from the motor, we scooted off down the ice. Behind came a crowd of small boys and men on their skates — most of them still doubtful I would fly. With a lurch and a mighty straining of wires we were in the air. It was amusing to look back and watch the skaters — they seemed to be going in every direction — bumping into each other

in their excitement at seeing a man actually fly. In taking off I had to clear one old Scot, so doubtful I would fly, that he had started off across the ice with his horse and sleigh. I think they both had the daylights scared out of them. I traveled three-quarters of a mile at a height of about 60 feet before again coming to the surface of the ice. I will say, the doubting ones overcame their feelings in short order.[49]

McCurdy's flight of February 23, 1909, qualified as the first by a British subject in the empire. Author Paul MacDougall of Sydney described the flight as "Silver wings over an arm of gold," referring to the aircraft over the literal English translation for the lake named Bras d'Or.

The excited Mr. Bell jumped up in his sleigh as the crowd cheered and threw hats and mitts into the air. The few who wagered in favour of a successful outcome did well. Even the seals in Baddeck, attracted by the engine and propeller noises, scrambled up on the ice in order to satisfy their curiosity.

McCurdy continued to glide the craft onto the ice. Seconds before touchdown, however, he noticed two young girls skating in front of his aircraft. Demonstrating his complete control of the machine, he gracefully steered to one side of the girls, avoiding a serious accident. The flight ended in a beautiful landing. "Flying the Silver Dart was just like being on a high after a couple of shots of whisky, I wanted to do it three or four more times,"[50] McCurdy said after completing his forty-eight-second flight.

However, Bell, mindful of the historic event

that had just occurred, restrained him, saying to the assembly gathered around McCurdy: "What we have seen just now may well prove to be one of the really important pages in history, I wouldn't want it to be spoiled. Douglas, you can fly her again tomorrow if you like, but that's all for today."[51]

After witnessing the historic flight, Mabel told her husband that the sight of the Silver Dart in the air made her more fully understand what had driven the other five members of the association to work so hard to build a successful flying machine. The people of Baddeck, and the world, had applauded.

Bell then invited everyone for some prepared raspberry vinegar cider and sandwiches at Beinn Bhreagh to celebrate. The majority of the villagers accepted and the festivities lasted into the night. After the flight, the wine cellars in the village of Baddeck reopened their doors, with free champagne for all. Many people in the village who had witnessed the flight signed a special book that Bell brought out to verify what had occurred. The names of 147 witnesses of McCurdy's first flight in the Silver Dart on Baddeck Bay are recorded in the proceedings of the AEA. The Silver Dart would soon enter the Canadian record books as the first powered, heavier-than-air machine to fly in Canada.

The experience of inventing the telephone taught Bell to be very aware of first-time events and how important it was to properly document such occasions. Bell telegraphed all the major newspapers in the US, Canada, Great Britain and France. Many of these papers carried the story in their editions the following morning. On February 24, 1909, Douglas McCurdy and the Silver Dart were back on the ice of Baddeck Bay. The Silver Dart again rose into the sky and reached slightly higher altitudes as McCurdy piloted Aeroplane No. 4 over four and one-half miles.

Reflecting on the event, Bell wrote,

*This may seem to be a small matter at the present moment; but when flying machines have become common and Aerial Locomotion a well-organized and established mode of transit, the origin and art in Canada will become a matter of great historical interest and people will look back to the first flight made on February 23, 1909, as the first flight of a flying machine in the Dominion of Canada.*[52]

The Silver Dart made fifty test flights — fourteen in Hammondsport, thirty-two in Baddeck and four in Petawawa — before it finally crashed. Pioneering aviation experts throughout the world were now producing superior aircraft, on par with or exceeding the Wright brothers' creations.

# 8 THE SILVER DART, A BETTER AIRPLANE

With each AEA airplane being built better than the preceding aircraft, at least seven improvements over the Wright's Flyer aircraft were built into the Silver Dart design.

The Silver Dart was built "like a watch,"[53] and it represented the culmination of the AEA efforts. The Dart embodied all of the advancements found in the previous three AEA airplanes, most notably the aileron for improved turning compared to the Wright's wing-warping method. As aileron design became more sophisticated and efficient, the flight control device entered into widespread use starting in 1915.

The front elevator was operated by means of a cantilever attached to a bamboo pole, which in turn was attached to the steering wheel (or yoke). Replacing levers with a steering wheel patterned after the automobile made controlling the aircraft easier to understand and has remained standard with companies like Boeing. Pulling back on the steering wheel operated the canard surface (the foreplane), causing the plane to climb; pushing forward would cause it to descend. On the original Silver Dart, rotating the steering wheel activated the rudder to help with turns (there were no rudder pedals). Wilbur and Orville devised slightly different flight lever controls in their Model "A" airplanes, with the Wright Flyer being available in either the "Wilbur Method" or the "Orville Method" control system. On May 14, 1908, after flying solo for seven minutes, Wilbur suffered his worst crash when — still not well-acquainted with the two new control levers — he apparently moved one the wrong way and slammed the aircraft into the sand at between 40 and 50 miles per hour (64 to 80 kilometres per hour). Luckily for him, he emerged with only bruises and a cut nose; however, the accident ended the practice flights and destroyed that airplane. According to General Henry ("Hap") Arnold, who learned to fly at a Wright school, "No two types of controls were the same in those days, and from the student's point of view the Wright system was the most difficult."54 The AEA used a yoke that all their pilots were familiar with, and it is still the standard in most of today's aircraft, which minimizes pilot confusion with flight controls.

Courtesy of Curtiss, the Silver Dart featured a fifty-horsepower, eight-cylinder, water-cooled engine. Due to more power being available from the Curtiss engines, AEA pilots always were able to sit upright like today's pilots do. The early Wright Flyer aircraft had less powerful four-cylinder engines, resulting in the pilot having to fly the aircraft lying on their stomach in an effort to reduce wind resistance (drag) to enable the takeoff. Eventually the Wrights adopted the sitting-up design with the model "A," but the extra engine horsepower gave the Silver Dart a superior performance early on, and AEA pilots never had to lie down to improve their aircraft takeoff performance.

The tricycle wheeled landing gear reduced friction resistance on the takeoff roll and was quickly adaptable under changing wind directions, compared

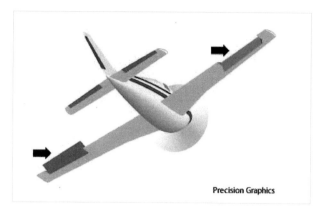

*On the rear of an airplane's wings, ailerons work in pairs to assist in turning.*

to the time-consuming laying of launching rails the Wrights used. The tricycle undercarriage, which became a standard feature on AEA aircraft, was made of steel tubing using motorcycle wheels from the Curtiss motorcycle factory.

The Silver Dart can be distinguished by its anhedral (downward-sloping) upper wings and dihedral (upward-sloping) lower wings. The result was a double-wing configuration that made the plane look like a person's narrowing eye when viewed head-on. The aerodynamic and stabilizing qualities of a dihedral angle were described in an influential 1810 article55 by Sir George Cayley. Angling the wings up or down spanwise from root to tip can help to resolve issues related to stability and control in flight. Bell had already detected the stability benefits of this design when choosing the dihedral, V-shaped angles for his tetrahedral kites.

- Dihedral: the wing tips are higher than the wing root, like an eagle's or hawk's wings. This design adds lateral stability and keeps the

*Front view of the airborne Silver Dart over Baddeck Bay with Kidston Island and Washabuck Mountain in the background.*

aircraft locked when it leans or banks into a turn, preventing it from slipping sideways.

- Anhedral: the wing tips are lower than the wing root, giving a "seagull" appearance. As the opposite of dihedral, it is used to increase manoeuvrability where other features result in too much stability, and it gives an aircraft more vertical control.

The Wright brothers wanted the pilot to have absolute control, and for that reason, their early designs made no concessions toward built-in lateral stability using dihedral wings. The Wrights designed their first powered 1903 Flyer with anhedral (drooping) wings, which are inherently unstable, but less susceptible to upset by gusty crosswinds — a good feature to have when operating off the windy shorelines of the Atlantic Ocean. At Kitty Hawk, Wilbur had to run alongside the Flyer to keep the wing tip from dragging in the sand.

The newer AEA wing design differed from contemporary French aircraft, starting with the AEA's Red Wing and carried on through to the Silver Dart. Known as the bowstring truss design, it was a slight variation of the Pratt design used earlier on the AEA Hammondsport Glider. J.H. Parkin wrote,

*In the bowstring truss, the two main wings, in frontal aspects, were bowed so that the wing tips were closer together than the centre sections of the wings, [creating the narrowing eye look]. The bowstring truss was introduced by Baldwin, primarily to provide a light structure, but it possessed also certain aerodynamic advantages . . . it was thought that the arrangement reduced the tilting effect of side gusts.*[56]

The combination of a dihedral and anhedral wing design gave the aileroned Silver Dart control that rivalled the Wright Flyer, making the Silver Dart's wings superior. The aerodynamics of this configuration were not well understood in 1909, certainly not by the courts that heard the Wright patent suit. A better understanding might have vindicated the AEA design as an alternative means of airplane control, putting an end to the litigation that ultimately hurt the Wrights.

Designed with just one propeller, the Silver Dart was a centreline thrust aircraft, which is easier to fly straight in the event of propeller failure. Bell and his team didn't borrow the idea of using two propellers, because with a more powerful V-8 engine a second propeller to bite more air wasn't required. Two different "elliptical" propellers drove the Wright's Flyers — it had a single engine powering two propellers simultaneously, making it look more like a twin-engine airplane.

The army was looking into purchasing an airplane from the Wright brothers and decided to conduct tests of a Wright machine at Fort Myer. On September 17, 1908, Orville Wright had no choice but to take AEA team member Lieutenant Selfridge to fly as a US military observer. That same day, Bell and Baldwin departed Baddeck for Washington to meet with lawyers about possible patents on the June Bug. They had only gone as far as Grand Narrows in Cape Breton when they received a telegram from Mabel with tragic news — Lieutenant Selfridge had been killed as a passenger in the crash of Orville Wright's aircraft. The flight began normally and uneventfully as Orville Wright completed four circuits of the Fort Myer test area. Halfway

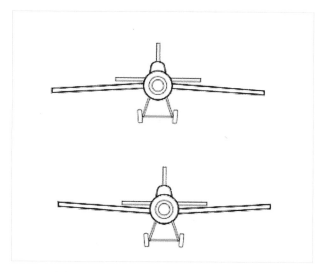

*Downward (anhedral) and upward (dihedral) sloped wings were combined to give AEA airplanes Baldwin's bowstring truss design.*

through the fifth circuit, flying at 150 feet (46 metres), the right propeller on the twin propeller-single engine pusher aircraft broke loose. The shattering propeller hit a guy wire bracing in the rear vertical rudder and the Flyer went into a nose dive. The craft hit nose first, burying the pilot and passenger in a twisted mess of wood, wire and cloth.

Pioneering pilots didn't understand or have experience yet in dealing with the consequences of asymmetric thrust or unwanted pull (yaw) toward the failed propeller, which occurs because the remaining good propeller is still producing full thrust. In the Wrights' crash in Virginia, the right propeller broke and the aircraft kept turning uncontrollably to the right in circles. This could never occur in the single engine and single propeller Silver Dart, making it a less complex aircraft and superior design.

Bell's thoughts on the Fort Myer disaster were that

double propellers introduce an element of danger. He thought it might be safer to use concentric propellers either pushing or pulling in a central line. Bell instructed McCurdy to not design the Silver Dart with two propellers and recommended the aviation community study the causes of the accident to avoid a repeat in the future.

Both men were taken to the Fort Myer hospital. Late in the evening, a doctor announced that Orville Wright was severely injured but he would live. The less fortunate Lieutenant Selfridge had struck his head on a wooden upright beam of the framework, fracturing his skull. He died later that evening on the operating table without ever regaining consciousness, and in doing so he became the first known passenger victim to die as a result of an accident in a powered aircraft. Selfridge was twenty-six years of age and he was buried at Arlington National Cemetery with full military honours. Selfridge's death cast a major pall over the members of the AEA. Mabel Bell in particular grieved him and afterward wrote a lengthy letter about her affection for Selfridge. The letter is a tribute to Selfridge's character and his abilities as an aerial experimenter.

With Selfridge's death, much of the sheer fun drained from the AEA. It was a grim reminder that they were not involved in child's play. The only consolation was the fact that Orville survived and the tragedy did not occur in an AEA airplane.

Based on eye witness reports of the Wrights' crash, Bell felt that Orville Wright had raised the front control too much, or too quickly, causing the head (aircraft nose) to rise, which caused the machine to lose its headway and stall. Shortly after the Fort Myer crash, Bell put forward another advanced notion of aviation:

> *Should not the front control be at the rear instead of the front? . . . Would it not be better to use a horizontal tail at the rear? The natural action of the wind of advance upon the front control is to upset the whole machine upwards or downwards as to make a complete somersault . . . Whereas the natural action upon a horizontal tail at the rear is to keep the longitudinal axis of the machine parallel to the line of advance and prevent any deviation up or down excepting by the will of the operator.*[44]

Bell had never gone up in an airplane and never would; in spite of lacking flying experience, he sensed that the long "heads" carried by practically all early airplanes were impractical as well as dangerous.

Many attempts to get airborne earlier failed because wing fabrics were often too porous. The new fabric used on the Silver Dart was a rubber and graphite-based silk, which had a silvery appearance and provided inspiration for the aircraft's name. Two important features of the Silver Dart's new fabric were that it could be glued instead of stitched and that it had low air resistance. This fabric was the best covering for the wings of any used by the AEA; it was impervious to air and thus maximized the wings' lift.

Building a superior aircraft alone, however, wasn't enough to succeed in 1909. There were unforeseen drastic changes brewing in the winds over AEA headquarters in Baddeck.

# 9 THE AEA DISSOLVES

As the AEA was reaching its agreed March 31 deadline, Bell and Mabel gave a great deal of thought as to how the work of the association could continue. Bell outlined his ideas of incorporating a joint stock company to manufacture airplanes and suggested that, since Curtiss had the most extensive business experience, he should be its manager. Strained relations arose between Curtiss and the association's other members, however, after Curtiss unexpectedly formed a new commercial aircraft building venture, the Herring-Curtiss Company, in March 1909. A request for him to attend the association's meeting and resolve the issue went unanswered.

Augustus Moore Herring was the son of a wealthy American cotton broker who built a glider in 1893 to get some aviation experience. Later he was hired by Octave Chanute to assist with experiments, and in 1895 Samuel Langley hired him to assist in his experiments, extending Herring's aviation experience. In 1909, Herring joined Curtiss to create the Herring-Curtiss Company before Curtiss had even resigned from Bell's team. It came as a great shock to Mabel and the other associates to learn that the Herring-Curtiss Company had formed to manufacture heavier-than-air machines before they could have even had a chance to try to extend the life of the AEA.

*Glenn Curtiss prepares for a flight in the June Bug at Stony Brook Farm in Pleasant Valley, Hammondsport, New York, in June 1908.*

At the last AEA meeting at Beinn Bhreagh Hall, Bell, Mabel, Baldwin and McCurdy were present. The last action directed to Mabel was a vote of "high appreciation of her loving sympathetic devotion without which the association would have come to naught."[57] In a solemn ceremony, the remaining members voted and at the stroke of midnight on March 31, 1909, the AEA dissolved.

Bell's team continued in their aerial pursuits after gaining experience with the AEA. Ultimately, Glenn Curtiss encountered the greatest success designing and building aircraft. He became known as the founder of the American aircraft industry, and his home in Hammondsport was known as the cradle of aviation. In 1911, Curtiss created the Curtiss Aeroplane Company, which designed and built aircraft for the US Navy. Long after the collapse of the AEA, Mabel would say,

*Those biplanes you saw flying overhead and thought of as Curtiss biplanes are not really*

*Curtiss biplanes at all, they are Baldwin biplanes constructed on AEA principles . . . The world does not know this — but Curtiss does, and it is due to him that the world does not . . . The other associates were mere enthusiastic boys, totally inexperienced in business, absolutely trustful of him. They were full of eagerness, fresh from college, but with no definite plans or prospects. None of them, or Mr. Bell, could have conceived that Curtiss was already scheming to betray them all in such a way as to make it practically impossible for them to reap their fair share of their mutual labours.*[58]

Mabel continued,

*After the association was dissolved, Casey and Douglas were thrown out of work . . . Very different was Glenn Curtiss' position. The close of the Association found him in sole possession of its workshop, which had been especially equipped for the making of aeroplanes; with men trained to the work and all the expensive, unprofitable, but absolutely necessary, preliminary experimental work done, and without cost to himself, and by trained mechanical engineers, whose services he could never alone have obtained. No other firm in the country, except perhaps the Wrights, was in such a favourable position to grasp the great opportunity to inaugurate a new business with immense possibilities.*[59]

Mabel went on to write,

*Douglas . . . discredited by Curtiss accomplished nothing and finally was glad to accept a position with Curtiss as a pilot — a position that acknowledged in no way his former equality with Curtiss . . . Glenn Curtiss stands before the public, a conspicuous figure, a millionaire, a successful man, who has developed a wonderful new industry. Casey Baldwin does not. Mr. Bell's name is rarely mentioned in connection with Aviation . . .*[60]

In spite of the disappointing setback, the remaining members of the AEA moved forward with aviation plans in Baddeck. Bell's team did their best to move onward and upward in the pursuit of excellent flying machines.

McCurdy later established the first Canadian aviation school in Toronto, which was owned by Curtiss. The Curtiss Flying School ran at full capacity, training air force pilots between 1915 and 1917 at Canada's first aerodrome, the Long Branch Aerodrome. In 1917, McCurdy was grounded from flying due to vision problems, but he continued working in aviation in a managerial capacity.

McCurdy was also instrumental in setting up an aircraft manufacturing company located in Toronto, Canadian Aeroplanes Ltd., which built aircraft for the Royal Flying Corps during World War I. McCurdy went on in 1928 to create the Reid Aircraft Company in Montreal, where he served as its first president, and he later helmed the

Curtiss-Reid Aircraft Company after its merger. Working in aviation until 1946, he developed an interest in politics, and from 1947 to 1952 he served as lieutenant-governor of Nova Scotia.

Some of Bell's last words in 1922 were to "Stand by Casey" — an encouragement to his family to continue Baldwin's work. After Bell died in 1922, Baldwin directed Bell Laboratories for ten years, until he entered politics and represented Victoria County in the Nova Scotia Legislature from 1933 to 1937. He was instrumental in bringing about the creation of Cape Breton Highlands National Park. Baldwin died at age sixty-six at Beinn Bhreagh in 1948. In 1974, he was inducted into Canada's Aviation Hall of Fame, founded the year prior. In his honour, the Casey Baldwin Award is granted annually by the Canadian Aeronautics and Space Institute to the authors of the best paper published in the Canadian Aeronautics and Space Journal.

# 10 THE CANADIAN AERODROME COMPANY

The AEA had been preparing to sell their planes at one-fifth of the Wrights' price before it dissolved. With financial backing from the Bells, the two remaining members of Bell's team, Baldwin and McCurdy, formed the Canadian Aerodrome Company (CAC). They intended to manufacture flying machines at Beinn Bhreagh in an effort to create an indigenous aircraft industry. Formed shortly after the dissolution of the AEA, the CAC was the first commercial enterprise in the British Empire to design and manufacture aircraft. In their first year, the CAC built the Baddeck Nos. 1 and 2. With these airplanes and the Silver Dart, they hoped to demonstrate the importance of flying machines to the Canadian government and obtain contracts for their construction.

Technically, the Baddeck No. 1 was the first powered, fixed-wing airplane of complete Canadian design and construction to fly in Canada. The plane was completed in July 1909, capable of carrying two people in tandem configuration, much like a modern-day motorcycle. It was produced by McCurdy, Baldwin and workers from the Baddeck area, and after its completion the people of Baddeck were allowed to view the airplane before it was disassembled and shipped to Petawawa, Ontario. The Baddeck No. 1 was an improved version of the

*A busy day at Bell's kite house factory at Beinn Bhreagh as ten workers fit ribs into the modern wings of the Baddeck No. 1 on June 21, 1909.*

Silver Dart and was planned as the prototype for the Canadian Aerodrome Company. At the same time, they prepared the parts for a second aircraft, which became the Baddeck No. 2. Among the innovations made to the Baddeck No. 1 were double surfaced (top and bottom) wings to reduce drag. The rubberized silk balloon cloth of the Silver Dart was replaced with a No. 10 grade cloth used on sails for yachts, which reduced the weight of the wings while still being waterproof and resistant to rot. The Baddeck No. 1 also sported an improved engine and a radiator built to provide additional lift.

Although the No. 1 was almost a replica of the Silver Dart in appearance, modifications were made to the frame using hollowed metal tubes with modular connectors. These tubes offered several advantages: greater strength due to the depth of the ribs; the metal was more durable; two layers of cloth could completely enclose the ribs; and the space inside the hollow metal was a good hiding space for the horizontal diagonal wires. After some problems with the Curtiss engine, it was replaced with a forty-horsepower Kirkham automobile engine.

After the Silver Dart was destroyed in experiments at Petawawa on August 4, 1909, the Baddeck No. 1 was shipped westward. The Baddeck No. 1 was assembled in Petawawa by August 6, but Bell warned in a telegram that McCurdy and Baldwin were to take it slow the first day and avoid taking any risks. He said the main purpose was just to be able to show that the airplane could actually fly, then on the second day of testing they could show

*Kite house factory workers, July 1909, dismantle the Baddeck No. 1 prior to its shipment to Petawawa to conduct trial military flights.*

off some of their aeronautical prowess. The first flight of the Baddeck No. 1 took place on August 11, 1909. The attempt was successful, with the craft flying a distance of 328 feet (100 metres). Engine problems resulted in curtailing further flights that day. The next day the Baddeck No. 1 was ready, after a takeoff run of 600 feet (183 metres), McCurdy lifted the nose and the airplane climbed very gradually. Airborne for 300 feet (91 metres) and gaining speed, the aircraft suddenly nosed up, climbed to thirty feet, stalled and fell backward to the ground, landing hard on its tail section. The CAC immediately realized it had been a mistake not to test fly the airplane in the privacy of Baddeck first.

They returned to Baddeck after the crash and Baldwin considered the problem thoroughly as to why the Baddeck No. 1, so superior in most respects to the Silver Dart, did not fly as well. They estimated that the cambered front elevator suddenly increased in leverage and caused an immediate, unexpected, nose-up action on takeoff. Bell felt the crash was due to this novel device never being previously employed, and it was likely not properly adjusted. McCurdy, not accustomed to its greater lifting power compared to the Silver Dart flight characteristics, got caught off guard, having insufficient time to check the backward tilt of the airplane on takeoff. After Bell's analysis of the Baddeck No. 1 accident was confirmed, they reverted to a flat front elevator. As always, the successes of Bell's team were based on logical, step-by-step development.

McCurdy was one of the first pilots to realize he got colder at altitudes of only 3,000 feet because temperatures decrease as one ascends. The summertime demo for the Canadian army in Petawawa was in warmer, less dense air, producing less lift. Carrying more weight with a passenger in warmer summer temperatures, the aircraft could not climb as well as in colder temperatures, which pioneer pilots didn't take into account. After the loss of

*The Baddeck No. 1 testing at Petawawa.*

the AEA's Silver Dart and the major damage to the Baddeck No. 1, the No. 1 was returned to Bell's facilities in Cape Breton.

After repairs and modifications, including new coil springs on the undercarriage, the installation of "between-the-wings" ailerons, an additional biplane tail installed and replacing the cambered front elevators with flat ones, the aircraft continued to fly in a series of proving flights. Back in Beinn Bhreagh, McCurdy flew the Baddeck No. 1 on its first post-repair flight, flying 295 feet (90 metres) over the ice of Baddeck Bay on February 23, 1910 (a year after the Silver Dart first flew there). In June 1910, the modified Baddeck No. 1 was shipped to Montreal to participate in the Montreal Air Meet, the first of its type in Canada. McCurdy flew short flights in Montreal on June 27, 28 and 30. Unfortunately, the No. 1 crashed again on June 30, damaging the aircraft, and it had to be shipped back to Baddeck for more repairs.

Over the course of the summer and fall of 1909, the newer and much improved Baddeck No. 2 aircraft would enter service. Lacking ice on Baddeck Bay, and having not perfected the seaplane just yet, it was a time to fly the Baddeck No. 2 as a land-based aircraft on wheels. The search was on by Bell and his team to find the best farm fields in the area to continue warm-weather flight testing. They found a suitable location on the Baddeck Valley River shoreline, in a large area of rich farmland that was free of obstacles and close to a river. The river banks of this farm flooded every spring, but by the late summer and fall, after the hay had been cut, it was prime

*On July 9, 1909, Bell's invitation attracted curious residents of Baddeck, who crammed into the kite house factory to take a close look at the new Baddeck No. 1 before it was shipped to Petawawa.*

real estate for pioneer practice flights where the CAC would gain more flying experience and expand their overall aeronautical knowledge beyond just getting airborne. The CAC was first in many ways to take part in advanced aviation thinking.

# 11 THE BADDECK No. 2 AEROPLANE

The Baddeck No. 2 was the sister ship to the No. 1, except it had an extra biplane tail.

The No. 2 was completed on September 11, 1909, built by Baddeck workers employed at the CAC factory in Beinn Bhreagh. Two days later, it was towed on a barge to the new testing grounds at the Bentinck Farm. The barge sailed out of Baddeck Bay, around Nyanza and up Middle River toward the farm. It was followed on September 19 by the Ugly Duckling, Bell's personal houseboat which was originally a catamaran barge that was later converted to a house boat for Bell's personal use to observe the Baddeck No. 2 test flights conducted at the Bentinck Farm. The flight tests on the roughed-out field began on September 17 and continued into November. Most of the first attempts, done by McCurdy, produced insignificant hops and rough landings, causing damage that they repaired immediately in order to resume testing. Baldwin took the controls and was able to get the craft ten feet (three metres) above the ground for a half mile. Pioneer pilots, lacking experience, often found that one sure way to get the aircraft to come down quickly within limited landing space was to shut the engine off. After shutting the engine off to decrease the lift from the wings, the pilot would try to land gently on the ice or grass.

*Damaged Baddeck No. 2 at Bentinck Farm.*

Bell watched Baldwin fly the Baddeck No. 2 during a bad landing and later said,

> From my distant point of view she seemed to be flying well, but not keeping at an even distance from the ground. Every few moments it appeared that Baldwin was about to alight, when he apparently changed his mind and steered her up again. The machine responded and rose without having touched the ground at the lowest point of her path. This maneuver was executed two or three times before the aerodrome shed was reached; but even here she did not land, and Baldwin rose again into the air although an end of the Baddeck River lay just beyond, which it would be necessary to avoid by a turn. It looked as though he intended to make a circuit of the testing ground, but in the middle of his turn, when he was quite near the River, he evidently decided to land and shut off the power. The moment the engine stopped, the head of the machine went up in the air like a rearing horse, and the [airplane] made a bad landing, striking on her port wing and tail. Considerable damage was done to the running-gear, and some struts were broken. Baldwin was unhurt; but Baddeck No 2 will be laid up for repairs for a few days.[61]

For the first week at Bentinck Farm, the Baddeck No. 2 only made high-speed ground runs. Bell informed his pilots that they should wait until they had made a satisfactory flight before letting the newspapers know that they were doing anything. Not testing the Baddeck No. 1 before

*Bell observes the Baddeck No. 2 over the Bentinck Farm from his houseboat, the Ugly Duckling.*

showing it to the military was a great lesson learned for Bell's team.

The last week in September 1909 was a much more productive week for test flights. On Saturday, September 25, John McDermid drove to the testing ground at Big Baddeck, where he reported that several successful hops were made with Baddeck No. 2 that day. Confirming that same day's events, Mrs. K.E. Baldwin's notes were placed in the *Beinn Bhreagh Recorder*, "When the wind went down about 5 o'clock took machine out and made four flights of about 200 yards each from 2 to 8 feet in the air."[63]

There were no flights September 26 or 27, although Bell noted in the *Beinn Bhreagh Recorder* that crowds of people showed up at Big Baddeck on September 26, 27 and 28. After successful flights on September 25, more flying was sure to attract onlookers.

More short-hop flights by the Baddeck No. 2 occurred between September 28 and 30; however, three flights on September 30 were more significant, as Bell noted:

> *A third attempt was made, starting from the distant end of the field, with Baldwin as aviator. This time there was no doubt of her lifting. The machine, I should judge, was about ten feet clear of the ground when she passed me; and made a flight of, I should say, about half a mile or more altogether.*[64]

The Baddeck No. 2 surpassed anything accomplished by the Silver Dart the previous winter, completing over sixty flights, mostly with McCurdy at the controls, both at the Bentinck Farm and on ice, into December 1909. The Baddeck Nos. 1 and 2 were also the first airplanes in which Bell's team used automobile engines successfully.

# 12 UNAUTHORIZED PASSENGER

During my research for this book, local Baddeck seniors indicated that an unauthorized and unrecorded passenger flight took place over Bentinck Farm, and that a Cape Breton woman was robbed of her rightful place in the aviation history books. However, as Mabel Eleanor MacRae Nicholson (eighty-seven years of age) stated when I interviewed her on May 16, 2015, at her home in Cape Breton, there is an extraordinary reason why the history of the first woman to fly as a passenger in the British Empire was not recorded when it happened, and why Dolena MacLeod's name has remained relatively unknown considering the magnitude of her claim to fame: "Bell's Law."

"Bell's Law" is alluded to in the book *Curtiss, Pioneer of Flight*. In this book on Curtiss, Bell is quoted on the high risks associated with carrying a passenger and the additional danger of operating two propellers. After 2,000 spectators viewed the crash on September 17, 1908, in which Selfridge died, Bell directed pilots Curtiss and McCurdy to dismiss thoughts of carrying a passenger in the June Bug or Silver Dart:

Bell to Curtiss: Mr. G.H. Curtiss, Hammondsport, Washington, September 29, 1908:

*Men repair the broken wheels on the Baddeck No. 2 at the Bentinck Farm airfield on September 20, 1909. Mabel Nicholson, who lived on the farm, tells how the wheels got damaged in her story.*

*The temptation is strong to attempt to carry two men in the June Bug, or in the Silver Dart, because Orville Wright, Wilbur Wright and Farmam or Delagrange have done it. I do not want you, or any of you to attempt it. It has already been done by others and of course we know that we can do it too, but I do not think that we have any right to run unnecessary risks . . . Orville Wright, the most experienced aviator of the world, and probably the best, has lost poor Selfridge. Do not let us, with less experience, run any risks of this kind . . . Remember Selfridge.*[65]

The Silver Dart was originally intended to be a two-person airplane, with the passenger seated in tandem behind the pilot. With this in mind, the Silver Dart was also to have dual propellers. That was the idea before the Fort Myer tragedy; however, after Bell put his foot down in his telegram to Curtiss, the ideas of carrying a passenger and having twin propellers on the Silver Dart design were dismissed.

Despite lacking official recognition, the story of Dolena MacLeod is well known in the Baddeck area of Cape Breton: at age twenty-three, she flew as a passenger with Baldwin on one of the thirty-six flights over Bentinck Farm. The problem facing true validation of Dolly's story is that all eye witnesses are now deceased, and before they passed, no one had officially documented Dolly's flight. However, it is difficult to dismiss the undocumented event as folklore due to the abundance of secondary source information backing her story. It is indeed possible that not even Dolly herself understood the significance of such a flight, at the time.

*During Canada's Centennial year, 1967, Dolly MacLeod takes her second flight, in a modern Cessna 180 airplane. This second flight took place fifty-eight years after her first, on June 29 in rural Alberta.*

The farmer's field used in Baddeck to do the thirty-six documented CAC flights in the fall of 1909 is located on the north shore of the Baddeck River in the Big Farm area, originally belonging to Vice-Admiral William Bentinck during the early 1790s. In 1909, Bentinck Farm belonged to Lieutenant Colonel Alexander David MacRae, better known by his nickname, Sandy. His daughter, Mabel Eleanor MacRae Nicholson, was born on November 25, 1927, and she heard the story of how Dolly flew with Baldwin recounted repeatedly by her father to visitors at their farm homestead.

With the absence of ice on Baddeck Bay, and in a time before man-made runways, Bell's team selected the best farmland available to fly from. As a result of being raised on the Bentinck Farm, in her interview Mabel provided a wealth of information on the 1909 test flying that occurred there. The kind of information she provided can't be found in history books — it's the kind that could only come from someone who lived on the farm where Bell's team conducted their flights.

Her interview confirmed why the Baddeck No. 2 aircraft suffered damage many times during the test flights. Mabel's father told her that due to a hump and hollow in the middle of the field, the Baddeck No. 2 aircraft broke a number of lightweight, motorcycle-type wheels when hitting that location during the takeoff and landing roll; furthermore, it was Mabel who first mentioned that Bell paid $10 to rent the farm (equal to almost three hundred dollars now). She also mentioned that several MacRae family members lived on both sides of the Baddeck River Valley in 1909, and her father told her that two of the MacRae sisters (cousins to Mabel) saw Dolly get in the aircraft with Baldwin.

According to Mabel, in 1909 it was well known in the Baddeck area that Bell strictly forbade any passengers be carried on board his test aircraft. It makes complete sense that after Orville Wright's Fort Myer crash that killed Lieutenant Selfridge on September 17, 1908, Bell would insist on not carrying passengers in order to reduce the risk of another unnecessary death. Respecting Bell's wishes (except once), it also makes sense that the young pilots, Baldwin and McCurdy, couldn't ever talk about the quick flight Dolly took. Mabel claimed that the community remained quiet about the flight for years to avoid showing any disrespect toward Bell.

It's understandable that with so many flights

*Mabel Eleanor MacRae Nicholson was raised on the Bentinck Farm where the Baddeck No. 2 flew several test flights in the fall of 1909.*

being conducted in Baddeck by the CAC in 1909, the opportunity existed for someone to fly as a passenger. When Dolly flew, several witnesses — all of whom are now dead — saw Dolly get in the airplane that Baldwin flew at Bentinck Farm.

The story of Dolly's flight in 1909 has been reported in at least seven different newspaper articles between 1951 and 2016. These reports state that McCurdy was present for this historic flight, along with Dolly's husband, James MacLeod, and her two young children, who watched anxiously. There were also reports of three unknown visitors from Sydney who watched this historic flight.

Ed Smith, a *Post Record* staff writer, wrote on July 12, 1951, "Preceding congratulatory reports from Matthew MacLean, federal M. P. for North Cape Breton-Victoria and William Carroll, M. P. for Inverness-Richmond, Chairman Buchanan introduced Mrs. James A. MacLeod [Dolly], 64, who is credited with being the first woman in the British Empire to make an airplane flight." In the accompanying photo, Dolly was attending the Victoria County Centennial as she posed before a model of the Silver Dart in Baddeck.

Eleanor Huntington, a reporter with the *Cape Breton Post*, was probably the most established writer to ever detail Dolly's story. She wrote an article that appeared on February 19, 1959, called "Ambition Satisfied." In that article, Dolly recalled her flight with Baldwin from fifty years ago in an interview. When asked if she had enjoyed her flight, she replied enthusiastically "Did I ever!" and explained,

> *That afternoon there were several scattered groups watching Casey Baldwin flying around. My husband and two little children were with me, and several of my girlfriends, though the only two I can name at the moment were my cousins, the MacRae girls.* [Alexandrina and Margaret, according to the *Victoria Standard Newspaper*, August 25, 2008.] *When Mr. Baldwin brought the plane down close to where we were standing and asked if any of the women would care to go up, I didn't wait to think it over, just got on board. The flight did not last long, anyway. We circled around within a radius of about three miles and I was back on the ground before many of them realized that I had gone.*

Bell died in 1922 and Baldwin in 1948, but McCurdy lived until June 25, 1961. He was alive between 1951 and 1959, when Dolly was appearing in newspapers saying she flew. McCurdy did not dispute Dolly's claim.

Later in the 1960s, Jane Havens wrote an article in the *Brooks Bulletin* (an Alberta newspaper) on Thursday, June 29, 1967, titled: "First woman to ride in aeroplane, Mrs. MacLeod takes second passenger ride 58 years later in EID." Havens wrote,

> On a beautiful summer day in 1909, Dolly MacLeod and her husband strolled across the grass to watch Casey Baldwin preparing to fly. As they came closer, Baldwin patted the seat beside him. "Climb in and try it," he invited. McCurdy, who stood nearby, laughed and said, "Dolly doesn't dare." She remembers climbing the three steps. It was something like getting on a disk-harrow, she recalls. Not closed in but subject to the elements. Gingerly she seated herself and waved farewell to her anxious-looking husband and yelled above the din, "All right, Casey. Let's go."

Havens wrote that Dolly told her,

> I didn't have sense enough to be scared. The noise was terrific but I was as proud as punch. The aircraft was performing perfectly. We soared up 50 to 75 feet and flew about a mile around our farm. When we came down and the plane hit the ground, it was quite noticeable. I hopped out and not realizing the noise had suddenly ceased. I threw my arms around my husband and yelled, "We did it!"

The Nova Scotian section of the Halifax *Chronicle Herald* on October 30, 2016, covered a story by writer Charles Thompson titled "Dolly MacLeod of Baddeck was the first woman to fly in the British Empire." Wrote Thompson:

> There are many adages about time — time solves all problems, time is on your side, time and tide wait for no man . . . Well, in the case of Dolly MacKay MacLeod, time is the enemy, not an ally. Time is slowly clouding memories and pertinent facts are being relegated to the dustbin of history. That is a shame, as the facts need to be acknowledged and given their proper due. Opportunity has been lost and with each passing day, the chance to give proper attention to a significant event is fading.

It wasn't only Dolly that made the mistake of calling the aircraft she flew in, "The Silver Dart." All the other professional news reporters did it as well. Dolly was sixty-four years of age on her first interview and approximately seventy-three years old when she did an interview for the fiftieth anniversary of the Silver Dart. She still remembered flying, but which aircraft she flew in has been questioned. Due to records kept by Bell's team,

*Blair Archer Fraser is Dolly MacLeod's grandson; he lived with his grandmother for several years as a young boy and remembers her telling him that she did fly with Casey Baldwin.*

we know it was the Baddeck No. 2. that flew out of Bentinck Farm.

Thérèse Peltier is popularly believed to have been the first ever woman passenger in an airplane, having flown on July 8, 1908, in Turin, France, when she flew as a passenger with Léon Delagrange. On October 7, 1908, Edith Berg, the wife of the Wright brothers' European business agent, became the second woman in the world to fly as a passenger with Wilbur Wright when they flew at an altitude of thirty feet for two and a half minutes at Auvours, France. Dolly MacLeod was the first woman in the British Empire to fly, in 1909, and she holds the title of being the third woman in the world. After Thérèse Peltier, Edith Burg and Dolly MacLeod flew, Grace MacKenzie became the fourth woman in the world to fly as a passenger a year after Dolly, with pilot Count Jacques de Lesseps aboard a Bleriot aircraft on October 25, 1910, near Belmont, New York.

Mabel directed me to Dolly's grandson, Blair Archer Fraser of Middle River, born December 26, 1952.

Fraser claims his grandmother was known to be a daredevil and would have definitely gone flying with Baldwin. On November 4, 2015, in an effort to preserve the history of Dolly's flight, Fraser and Mabel went before lawyer Sandy Hudson in Baddeck, Nova Scotia, and swore in separate statutory declarations that Dolly did fly with Baldwin at Bentinck Farm in 1909, according to the secondary sources of information that they personally received (Fraser talked directly with his grandmother and Mabel spoke with her father).

# 13 ADVANCED THINKING

In 1892 Bell described the "flying machine of the future"[66] as having the same basic features as the helicopter — which wouldn't be developed for another forty years. Bell explored numerous rotor blade shapes and ingenious "winged flywheels" in his experiments, which often flew to heights of two hundred feet using either mechanical power or rockets with gunpowder. The following year, Bell declared, "I have not a shadow of doubt that the problem of aerial navigation will be solved within 10 years."[67]

Bell's vision of flight was as sweeping as his grand concept of the telephone; he foresaw the strategic importance of military air power when he wrote in a magazine in 1908, "The nation that secures control of the air will ultimately rule the world."[68] He also foresaw rockets flying into space when in 1915 he said,

> *I have no doubt that a machine will be driven from the Earth's surface at enormous velocities by a new method of propulsion. Think of tremendous energies locked up in explosives. What if we could utilize these in projectile flight?*[69]

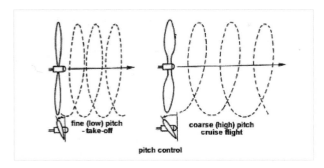

*Fixed pitch wooden propellers were limited in all phases of flight performance. Variable pitch (angles) propellers are common today.*

Baldwin foresaw the need for a variable pitch propeller long before it existed. A propeller screws into the air like a wood screws into a piece of wood, so think of it like an airscrew. Much like wood screws can have a fine or coarse thread, propellers can have a fine or coarse thread (known as pitch). The fine pitch is best suited for takeoff and landings when considerably more bites into the air are needed, occurring when the engine is at the highest revolutions per minute. When an aircraft is in level cruising flight, the engine is powered back to prolong engine life and save on fuel consumption. Using a coarse pitched propeller at this stage of flight the engine will only need to rotate at lower revolutions per minute because each single bite of air taken by the coarse pitch propeller is a much bigger bite into the air allowing the engine to run slower. He indicated the difficulty would be in designing a propeller that could change its pitch, allowing for a wide range of speeds that would be efficient for takeoff, landing and cruise speeds. The variable pitch propeller is common on modern aircraft.

Baldwin proposed that the usual methods of measuring speeds for an airplane were wrong unless they accounted for headwinds and tailwinds. If an aircraft were flying forty miles per hour against a ten mile per hour headwind, he thought that the aircraft would lose ten miles per hour, which would result in a decreased speed over the ground of just thirty miles per hour. The reverse is also true — if that same aircraft, still doing an airspeed of forty miles per hour, did a U-turn, gaining a ten mile per hour tailwind pushing the aircraft, the speed over the ground of the aircraft would increase up to fifty miles per hour. Bell agreed, and concluded that the

*Baldwin recognized the need for variable pitch propellers to replace the wood propellers.*

*Alexander Graham Bell sitting on the Dhonnas Beag with his hand on his head, deep in thought about better ideas on flying boats.*

mountain downdrafts that they experienced in the Baddeck River Valley were preventing the aircraft from climbing well after takeoff.

Baldwin estimated that at 16,000 feet the air was about one-half the density of the air at sea level, meaning the thin air at 16,000 feet would provide less wind resistance against the airframe, and he felt this would translate to much less fuel being required to take an aircraft across the Atlantic Ocean. Baldwin knew that aircraft would be capable of more speed facing less wind resistance at higher altitudes, so he advised to carry oxygen and fly faster. That's because above 10,000 feet, where the air starts to thin out considerably, the use of oxygen to support human respiration is necessary. Baldwin also suggested that the airplane should be enclosed to allow the pilot to be immersed in still air, remaining warm, because for every 1,000 feet an aircraft ascends the outside temperature drops approximately 2 degrees Celsius. This was a fact first realized by McCurdy, who did more flying than Baldwin.

Bell envisaged an eastbound Atlantic crossing at 25,000 feet, at a speed three times 40 miles per hour (the average speed of airplanes in 1909) plus a 30 mile per hour tailwind, would equal a speed of 150 miles per hour. He believed that in about thirteen hours you could cross the Atlantic Ocean. Bell said, "I believe that it will be possible in a very few years for a person to take his dinner in New York at 7 or 8 o'clock in the evening and eat his breakfast in Ireland or England the following morning."[71] In the *Beinn Bhreagh Recorder* of December 15, 1909, Baldwin concluded that if an airplane were to fly at higher altitudes, it must travel at a higher speed in less dense air, or how else can it be supported?

Baldwin continued his study of the altitude problem and in April 1910 presented Bell with graphs, similar to those used by pilots today, of the total horsepower to drive the Baddeck Nos. 1 and 2 at different air densities and altitudes. McCurdy too had his own vision that airplanes would be important not only in time of war, but that they could be used for the purpose of exploration in the far north, where explorers encountered so many hardships and privations. There is little doubt that Bell and his team were much more advanced in their thoughts than many other aviators at the time.

## Success with Water Landings

Water testing resumed on June 1, 1910, with McCurdy flying the Baddeck No. 2 which became the CAC's first and only amphibian airplane after a single toboggan type float was mounted between the main wheels under the centre section. Additional small, single stabilizing floats for balancing the aircraft (one on each side) were fitted under the inner struts of the lower wings. Getting flying boats to break free from the suction created by the water was always a problem for pioneer aviators and hence the original reason for the AEA hydrofoil experiments. Baldwin suggested that Long Sand Point at Beinn Bhreagh harbour be furnished with a road for a wheels takeoff followed by a water landing in Baddeck Bay. The Beinn Bhreagh Farm Department graded a road on the beach for a takeoff strip, however the unrolled road proved to be too soft (as in Petawawa) for the narrow bicycle type wheels. The airplane was transported across Baddeck Bay on a barge to Haliburton Beach where there was good solid turf and adequate takeoff distance available. Performing a wheels takeoff, the Baddeck No. 2 Aeroplane lifted off the beach and seconds afterwards landed straight ahead in shoal waters. Although pilot McCurdy was drenched by the resulting deluge, the engine and wings escaped wetting and remained undamaged. The airplane did not overturn and the test was considered a success — it was the first successful powered amphibious seaplane flight in the British Empire. The optimistic Bell said that there was no finer machine than the Baddeck No. 2, and that Baldwin and McCurdy had no need to worry over the future because their aircraft kept getting better.

Testing the Baddeck No. 2 dramatically increased the knowledge of this group of pioneer aviators and it is more evidence of their ability to think well into the future. Through methodical experimentation on each of the airplanes, the Bell team made huge leaps and advances in aviation knowledge.

The CAC would soon advance into the world of monoplanes by producing the first monoplane — named the Hubbard and nicknamed "the Mike" — in North America. After the monoplane, Bell continued to explore the use of the tetrahedral design.

# 14 THE FINAL THREE AEROPLANES

### The Oionus I

The Oionus was one of Bell's final fusions between kite and plane. It was based mostly on the Silver Dart, but still used tetrahedral cells. It was Canada's first and only triplane, designed by Bell with the help of McCurdy and Baldwin as consulting engineers. On March 2, 1910, they christened this flying machine the Oionus. In Greek, Oionós means omen.

The Oionus I incorporated tetrahedral sections in the wing structure of the triplane design, as had been used in Bell's AEA Cygnet series. The triplane employed a longer central plane with wing tip ailerons and flying controls, based on a fixed biplane tail. Like other AEA airplanes, it had a rudder at the rear and a canard biplane elevator section at the front. The internal structure was tubular steel with linen-covered wings and interior sections; a four-wheel chassis or running gear formed the undercarriage. A Curtiss pusher engine drove a propeller through a chain and sprocket arrangement; later, a Kirkham engine from the Baddeck No. 2 was substituted.

They tested the Oionus I between March 10 and March 25; however, none of the trials were successful. On March 25, 1910, a test flight off the ice at Baddeck Bay succeeded in only three of the

*Bell's Oionus I, tetrahedral cell airplane, being tested on Baddeck Bay, March 25, 1910.*

four wheels coming off the ice. Based on the fact that the Cygnet I flew, the Oionus probably would have flown with a larger motor; however, shortly afterward the ice melted and no further attempts were made to fly the Oionus I.

On March 12, Baldwin was conducting some high speed ground tests in the Oionus I just as McCurdy was making a landing in the Baddeck No. 2. To the spectators behind the Oionus, it looked as if a collision took place, and it was in fact Canada's first near-collision between two aircraft. Everyone was relieved when they realized that no one was hurt and the two planes had not actually collided.

## The Hubbard Monoplane (The Mike)

By April 5, 1910, the disappearing ice on Baddeck Bay left very little time for further trial flights with the Hubbard Monoplane. That day, though, "the Mike," as it was nicknamed, did nine successful flights over the ice, reaching elevations of approximately 15 feet,

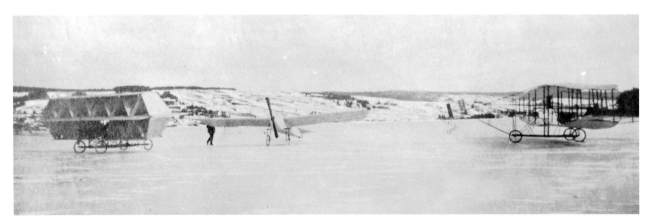

*Three CAC aircraft on Baddeck Bay with the Hubbard (Mike) Monoplane parked between the Oionus I and the Baddeck No. 2, as part of Bell's 1910 winter flying program.*

and as the third plane built by the CAC, the flights it flew on Baddeck Bay were the first witnessed flights of a monoplane in Canada. The plane was first piloted by McCurdy and later by Mabel Bell's father, Gardiner Green Hubbard, who had commissioned the plane. Hubbard took the plane home to Boston when he returned, which qualified the Mike as the first plane built in Canada for export.

Improved structural techniques, lighter building materials and the quest for greater speed made the biplane configuration obsolete for most purposes by the late 1930s, and the monoplane became the preferred choice. Later aircraft manufactured in Ontario by Curtiss's company are sometimes erroneously referred to as the first to be exported out of Canada; however, Bell's team exported the Mike first, out of Baddeck.

## The Cygnet III

The Cygnet, rebuilt as the Cygnet III with a more powerful seventy horsepower Gnome Gamma engine, flew a final flight on March 17, 1912, over the Bras d'Or Lake, piloted by McCurdy.

The results were highly unsatisfactory — the Cygnet III was only able to lift off and fly approximately two feet off the ground. Further flight attempts in the Cygnet III were abandoned after its March 17 flight, when the tetrahedral cell bank failed structurally and left the aircraft irreparably damaged. It seemed clear that this design was not going to yield a controllable aircraft and Bell's attempts at making a successful lightweight airplane from tetrahedrals had failed. Bell, now in his sixties, accepted disappointment with the tetrahedral design. It may have been Bell's final aviation pursuit, but Bell was a man who was ahead of his time in thoughts on aviation. His longstanding idea that a motorized kite should fly was not out of the question, although Bell's powered and manned tetrahedral aircraft did not fly well in his time.

Today there are many variations of powerful, lightweight, motorized kites that do fly well, such as the ultralight trike. The modern-day ultralight trike aircraft amounts to not much more than the flying kite that Bell visualized in the late 1800s, driven by an engine with a pusher-type propeller like the Silver Dart had. The ultralight trike pilot keeps their hands on a crossbar, which is used to shift the pilot's weight in order to steer. There are no moveable aileron wing tips or a tail on this kite-like aircraft. It is a very simple design and more proof that a motorized kite using weight-shift steering technology can fly well.

In June 1903, Bell wrote,

> *I have had the feeling that a properly constructed flying machine should be capable of being flown as a kite; and conversely, that a properly constructed kite should be capable of use as a flying-machine when driven by its own propellers.*[73]

The problem with Bell's tetrahedral kites turned out to be too much weight with too little engine power; however, his vision of a flying kite, like many of his other visions, was years ahead of his time.

# 15 THE CANADIAN AERODROME COMPANY STRUGGLES TO SURVIVE

In March 1909, Mr. Fred Cook, a special correspondent of the *London Times* at Ottawa, asked Bell how long they proposed to continue aviation experimental work. Bell, believing the question might possibly be an unofficial enquiry from the Governor General, Earl Grey, saw opportunity for the CAC to sell aircraft. Having no sons of his own, Bell felt great affection for the two remaining members of the AEA and focused on helping them establish a career in the infant industry of manufacturing aircraft. He encouraged Baldwin and McCurdy to continue to work together beyond the experimental stage, building aircraft for employment in warfare.

Developments in Germany, France and Italy, which by 1909 all employed the use of airplanes, generated public concern in Britain. With foreign advances in aviation, Bell realized the British government was alarmed at the activity of foreign nations in developing aviation for war purposes. Major G.S. Maunsell, the Director of Engineering Services, requested a leave in the US to study developments in military engineering. Maunsell, Canada's

*Bell and Mabel walking through the grounds of Beinn Bhreagh.*

first advocate for military aviation, reported to his superior, Colonel R.W. Rutherford, the Master-General of the Ordnance for Canada. Rutherford proposed to the Militia Council in March 1909 that the department adopt an aviation policy. Questions in the Canadian Parliament indicated the government was alive to the fact the CAC was leading in aviation development. To Bell, everything indicated governments may be willing to give aid toward airplane manufacturing.

Interest by Great Britain and Canada culminated into an invitation for Bell to address the Canadian Club of Ottawa and meet the cabinet after successful flights in Baddeck aroused a strong feeling of patriotic pride among Canadians. Bell agreed to a private meeting with the governor general, Prime Minister Sir Wilfrid Laurier and his cabinet. Speaking to the Canadian Club on March 27, 1909, about his group's successes, Bell suggested that Canada secure the services of the CAC.

In April 1909, St. Petersburg requested the CAC undertake building airplanes for Russia. Bell had McCurdy write to Minister of Finance William Stevens Fielding immediately to pressure the Canadian government and to let him know — and through him the Canadian cabinet and the representative of Great Britain — that inquiries had been received from Russia. Canadian Deputy Minister of Militia and Defence Colonel Eugène Fiset wrote to Bell in May 1909, stating he regretted that no funds for airplane construction had been provided by Parliament for the present year. "No funds provided by Parliament" is a much-used tactic of political legerdemain to evade responsibility. To show some interest, the Militia Council, acting on the Master-General of the Ordnance Rutherford's recommendation, invited the CAC to conduct flying trials at Petawawa.

Of particular interest to the media was the disclosure that an additional new airplane, the Baddeck No. 1, would also be flown. An official party from Ottawa arrived at Petawawa in early August, including Deputy Minister of Militia and Defence

*Baddeck No. 2 on March 11, 1910, over the ice of Baddeck Bay, performing a demonstration flight for Major Maunsell.*

Colonel Fiset, Master-General of the Ordnance Colonel Rutherford, Director of Engineering Services Major Maunsell and Major-General W.D. Otter. Four times on August 2, 1909, McCurdy flew the Silver Dart a half mile at a height of ten feet, alternating Baldwin and workman McDonald as passengers. The fourth flight proved to be the Silver Dart's last flight; as McCurdy prepared to land, he was blinded by the rising sun. The front wheel of the tricycle undercarriage struck the edge of a sandy knoll, causing the aircraft to porpoise and crash on its starboard wing. The centre section and the elevators were shattered and the wings damaged, leaving only the engine intact. Baldwin and McCurdy escaped with minor injuries. The political brass ordered more trials and the Baddeck No. 1, designed to be Canada's first military aircraft, was unpacked.

On August 11, 1909, flight trials of the similar looking Baddeck No. 1 continued, with the first flight being successful, but unfortunately, the second flight on the next day crashed. McCurdy was only slightly bruised, but the machine considerably damaged. Holding Petawawa flight trials on an undulating cavalry field full of ridges, knolls and sandy terrain proved to be quite unfit for the narrow tires of these early vintage machines, yet in spite of the handicap, four of the six flights succeeded.

Several Canadian MPs and Militia highbrows were unimpressed; however, Rutherford continued to foresee a valuable scouting role. Colonel

Eugène Fiset, a veteran who fought in battle, was impressed by McCurdy's courage. Deputy Minister Major-General W.D. Otter thought the airplane was too expensive a luxury for Canada. Questions regarding the government's aviation policy continued to be raised in the House of Commons in November 1909 with Sir Frederick Borden indicating the policy was guided by the actions of the War Office in England.

Governor General Earl Grey was impressed by December 1909 flights in Baddeck, flown by McCurdy in the Baddeck No. 2. Grey, flushed with imperial pride, praised the CAC efforts, telling Bell after personally witnessing the flights, it enabled him to realize that the flying machine would make 'Dreadnoughts' as obsolete as bows and arrows.

On March 3, 1910, the CAC wired the Militia Council, requesting Maunsell spend time in Baddeck observing flights. The flat ice on Baddeck Bay provided a superior runway, delivering enhanced aeronautical performances. The CAC made several good flights during Maunsell's stay, including one of six-and-a-half minutes. McCurdy demonstrated the capability of the machine to carry two persons, taking Major Maunsell for two flights.

## The Offer

The CAC informed the Militia Council on March 10, 1910, that they respectfully offered the Canadian government two airplanes, the repaired Baddeck No. 1 and the new Baddeck No. 2, for the total sum of $10,000, and, at no extra expense, they offered to train selected army officers. Rutherford was aware the attitude of the British War Office was to encourage experiments, but to buy nothing. Concerned about losing the expertise possessed by the CAC, Maunsell called for the department to offer a grant of $10,000, to encourage them to continue their research. His superior, Rutherford, favoured and summarized Maunsell's report, passing it to the Militia Council for consideration.

The report was passed to the cabinet on April 7, 1910, and the application rejected. With the CAC anxious for sales, Colonel Rutherford went back to Colonel Fiset, asking whether a grant of $5,000 could be made from the funds voted for the Engineer Services. A second rejection was received in early June.

Continuing the fight, Major Maunsell again recommended on August 1910 that $10,000 be included in the 1911–12 estimates as a grant to the now defunct CAC (assuming the CAC would be willing to resume after closing in April). The Militia Council, rejected the proposal a third time on September 13, 1910. In late October 1911, in a joint report to the minister, Maunsell and Fiset stated that it would be preferable for the department to secure two airplanes and train aviators, after the Italians already proved airplanes valuable in war reconnaissance work.

Money became available in the budget in 1912 for the purchase of aircraft and training. Major-General Sir Colin John Mackenzie, the new Chief of the General Staff, sought authority for a start on a modest aviation program suggested by the War Office in answer to a Canadian request by Maunsell.

Colonel Sam Hughes, now Minister of Militia and Defence, rejected the proposal a fourth time. Placing the final nail in the aviation coffin, Hughes prevented the resurrection of the CAC. Disappointed, Deputy Minister Fiset reported Hughes did not want any steps taken this year — neither toward training nor purchase of airplanes. Sam Hughes blustered at McCurdy: "The aeroplane is an invention of the devil and will never play any part in such a serious business as the defence of the nation, my boy!"[76]

## Bell's Legacy

Bell was influential in improving airplane design, bringing airplanes of his day close to modern-day standards. His monumental effort in producing airworthy aircraft is recognized today. In the Bell Museum, a proud replica of the famous Silver Dart is silently suspended from the ceiling, symbolic of Bell's impressive aviation career in that community.

As in 1885, when the Bells first arrived in Baddeck, the village remains a small community and the stories of biplanes buzzing over Baddeck Bay and the Bentinck Farm are just memories of a faded past.

Aviation truly was Bell's highest calling.

After World War I, at the American government's request, Bell studied and outlined a policy for military aeronautics. He said,

> Air power will prove to be the decisive factor in future wars . . . We may conclude that neither our Army nor our Navy can defend the United States from attack through the air. This requires the addition of a third arm to our system of national defense, a National Air Force, quite distinct from the Army and Navy, capable of cooperating with both, and also capable of acting independently of either. Also, it should be provided with a special college upon the model of those at West Point and Annapolis.[22]

These prophetic ideas would not be adopted until after World War II. Though his legacy in aeronautics may never be recognized as having the same magnitude as his other accomplishments, he without a doubt left an indelible mark upon aviation, not only through his own work, but by stimulating interest in aeronautical research by others.

## Parallel Flight Paths

Fifty years after the CAC spiralled out of the sky, on February 20, 1959, Conservative Prime Minister John Diefenbaker announced the cancellation of the Avro Canada CF-105 Arrow. Avro Canada's highly skilled personnel scattered to the US, working on the Apollo Missions, and Canadians again bemoaned the devastation of Canada's aircraft industry.

# Endnotes

1. Lilias M. Toward, *Mabel Bell, Alexander's Silent Partner* (Wreck Cove, NS: Breton Books, 1996), 235, 255.
2. "Beinn Bhreagh, Nova Scotia," *Wikipedia*, last modified May 14, 2017, http://wikivisually.com/wiki/Beinn_Bhreagh.
3. Judith Tulloch, *The Bell Family in Baddeck* (Halifax: Formac Publishing, 2006), 75.
4. Naomi Pasachoff, *Alexander Graham Bell, Making Connections* (New York: Oxford University Press, 1996), 118.
5. M. Robinson, "Alexander Graham Bell's Flights of Fancy," *Kiting Magazine*, 23, no. 4 (Fall 2001), http://www.kitehistory.com/aeolus/Alexander_Graham_Bell.html.
6. Cecil R. Roseberry, *Glenn Curtiss, Pioneer of Flight* (New York: Syracuse University Press, 1991), 69.
7. Drachen Foundation, "Meet the Kite Maker: Alexander Graham Bell," 2006, www.drachen.org/sites/default/files/pdf/MKM-AG-Bell.pdf.
8. Mary Kay Carson, *Alexander Graham Bell: Giving Voice to the World* (New York/London: Sterling Publishing, 2007), 102.
9. Robert V. Bruce, *Bell: Alexander Graham Bell and the Conquest of Solitude* (New York: Cornell University Press, 1990), 432.
10. M. Robinson, "Alexander Graham Bell's Flights of Fancy," *Kiting Magazine*, 23, no. 4 (Fall 2001), http://www.kitehistory.com/aeolus/Alexander_Graham_Bell.html.
11. Ibid.
12. "Angle of Climb," *Wikipedia*, last modified April 12, 2017, https://en.wikipedia.org/wiki/Angle_of_climb.
13. M. Robinson, "Alexander Graham Bell's Flights of Fancy," *Kiting Magazine*, 23, no. 4 (Fall 2001), http://www.kitehistory.com/aeolus/Alexander_Graham_Bell.html.
14. "Bell, Alexander Graham," *National Aviation Hall of Fame*, 2017. http://www.nationalaviation.org/our-enshrinees/bell-alexander/.
15. Cecil R. Roseberry, *Glenn Curtiss, Pioneer of Flight* (New York: Syracuse University Press, 1991), 13.
16. Alexander Graham Bell, "The Flying-Machine of the Future," *Beinn Bhreagh Recorder*, June 5, 1892, as dictated to Arthur W. McCurdy.
17. Dorthy Eber, *Genius at Work: Images of Alexander Graham Bell* (New York: Viking Press, 1982).
18. Lilias M. Toward, *Mabel Bell, Alexander's Silent Partner* (Wreck Cove, NS: Breton Books, 1996), 170.
19. Octave Chanute, "Progress in Flying Machines," *The American Engineer and Railroad Journal*, 1894, 219.
20. Cecil R. Roseberry, *Glenn Curtiss, Pioneer of Flight* (New York: Syracuse University Press, 1991), 74.
21. Ibid., 75.
22. "Bell, Alexander Graham," *National Aviation Hall of Fame*, 2017. http://www.nationalaviation.org/our-enshrinees/bell-alexander/.
23. John Boileau, "A Century Aloft: The Rise of the Silver Dart," *Legion Magazine*, February 9, 2009, https://legionmagazine.com/en/2009/02/a-century-aloft-the-rise-of-the-silver-dart/.
24. J.H. Parkin, *Bell and Baldwin: Their Development of Aerodromes & Hydrodromes at Baddeck, Nova Scotia* (Toronto: University of Toronto Press, 1964), p. vi.
25. John Boileau, "A Century Aloft: The Rise of the Silver Dart," *Legion Magazine*, February 9, 2009, https://legionmagazine.com/en/2009/02/a-century-aloft-the-rise-of-the-silver-dart/.
26. Rannie Gillis, "Dynamic Four-Man Team Work Tirelessly to Build Flying Machines," *Cape Breton Post*, October 6, 2008.
27. Ibid.
28. "Alexander Graham Bell: His Aeronautical Experiments," *Carnet de Vol*, accessed January 30, 2017, https://www.carnetdevol.org/Bell/aeronautical.html.
29. John Boileau, "A Century Aloft: The Rise of the Silver Dart," *Legion Magazine*, February 9, 2009, https://legionmagazine.com/en/2009/02/a-century-aloft-the-rise-of-the-silver-dart/.
30. Lilias M. Toward, *Mabel Bell, Alexander's Silent Partner* (Wreck Cove, NS: Breton Books, 1996), 179.
31. "Otto Lilienthal — 456th F.I.S.," *The 456th Fighter Interceptor Squadron*, accessed February 1, 2017, www.456fis.org/THE_HISTORY_OF_FLIGHT_-_OTTO_LILIENTHAL.htm.
32. "Selfridge Aerodrome Sails Steadily for 319 Feet. At 25 to 30 miles an Hour," *Washington Post*, May 13, 1908.
33. Ibid.
34. Robert Esnault-Pelterie, "Expériences d'Aviation, Exécutées en 1904, en Vérification de Celles des Frères Wright," *L'Aérophile*, June 1905, 132–138, https://archive.org/stream/larophile13besa#page/132/mode/2up.
35. There is some controversial history over the invention of the aileron and wing warping, with seven claims to be the first use. The claims included: 1868, Matthew Boulton of England (aileron); 1897, Louis Mouillard of France and Octave Chanute (wing warping); 1903, Richard Pearse of New Zealand (aileron); 1903, Wright Brothers (wing warping); 1909, Henri Farman of France (aileron); 1911, Aerial Experiment Association (aileron); 1913, Glenn Curtiss (improvements on the AEA aileron).
36. "The Flight Of The 'June Bug,' — 456th F.I.S." *The 456th Fighter Interceptor Squadron*, accessed February 1, 2017, http://www.456fis.org/THE_JUNE_BUG_-CURTISS.htm.

37. Ibid.
38. Tony Foster, *The Sound and the Silence: The Private Lives of Mabel and Alexander Graham Bell* (Lincoln, NE: iUniverse, 2000), 319.
39. Ibid., 320.
40. "Bell, Alexander Graham," *National Aviation Hall of Fame*, 2017. http://www.nationalaviation.org/our-enshrinees/bell-alexander/.
41. Octave Chanute, "Aeroplanes: Part XV, September 1893," in *Progress in Flying Machines* (1893), http://invention.psychology.msstate.edu/inventors/i/Chanute/library/Prog_Aero_Sep1893.html.
42. J.H. Parkin, *Bell and Baldwin: Their Development of Aerodromes & Hydrodromes at Baddeck, Nova Scotia* (Toronto: University of Toronto Press, 1964), 98–99.
43. Hydro- a combining form meaning water, becomes hydroplane, much like aero- a word-forming element meaning air, becomes aeroplane.
44. Cecil R. Roseberry, *Glenn Curtiss, Pioneer of Flight* (New York: Syracuse University Press, 1991), 141.
45. Alexander Graham Bell et. al., "Drome No. 6," *Bulletins of the Aerial Experiment Associaton*, no. XVII (November 5, 1908), 7, http://www.loc.gov/resource/magbell.14100102/?sp=208.
46. Alexander Graham Bell et. al., *Bulletins of the Aerial Experiment Associaton*, no. XVIII (October 29, 1908), 32, http://www.loc.gov/resource/magbell.14100102/?sp=233.
47. Mabel Bell to Daisy Fairchild, February 23, 1909. Alexander Graham Bell Family Papers, Library of Congress, https://www.loc.gov/resource/magbell.14800303/?st=text.
48. Fraser's Pond is located near the Bell Museum. When exiting the museum driveway, turn left and drive 2.6 kilometres northeast toward Sydney to find Fraser's Pond on the left.
49. "Members of the AEA," *Best Breezes*, last modified February 3, 2017, http://best-breezes.squarespace.com/members-of-the-aea/.
50. John Boileau, "A Century Aloft: The Rise of the Silver Dart," *Legion Magazine*, February 9, 2009, https://legionmagazine.com/en/2009/02/a-century-aloft-the-rise-of-the-silver-dart/.
51. "The Silver Dart — Aerodrome 4," *Best Breezes*, last modified February 3, 2017, http://best-breezes.squarespace.com/the-silver-dart-aerodrome-4/.
52. "AEA Silver Dart — Full Size Replica" *Hangar Flight Museum*, http://www.thehangarmuseum.ca/exhibits/aea-silver-dart-full-size-replica.
53. "The Flight Of The 'June Bug,' — 456th F.I.S." *The 456th Fighter Interceptor Squadron*, accessed February 1, 2017, http://www.456fis.org/THE_JUNE_BUG_-CURTISS.htm.
54. Phaedra Hise, "How the Wright Brothers Blew It," *Forbes*, November 19, 2003, http://www.forbes.com/2003/11/19/1119aviation.html.
55. Charles H. Gibbs-Smith, *Sir George Cayley's Aeronautics 1796–1855* (London, H.M. Stationery Off., 1962), 223.
56. J.H. Parkin, *Bell and Baldwin: Their Development of Aerodromes & Hydrodromes at Baddeck, Nova Scotia* (Toronto: University of Toronto Press, 1964), 145.
57. *Aerial Experiment Association Chronology*, p. 6, accessed January 31, 2017, studylib.net/doc/8476451/aerial-experiment-association-chronology.
58. Lilias M. Toward, *Mabel Bell, Alexander's Silent Partner* (Wreck Cove, NS: Breton Books, 1996), 214.
59. Ibid., 215.
60. Ibid., 216.
61. Alexander Graham Bell, "Third Flight Sept. 30," *Beinn Bhreagh Recorder*, September 30, 1909, p. 254, https://memory.loc.gov/mss/magbell/307/30700101/30700101.pdf.
62. Alexander Graham Bell, "Cause of Accident by AG Bell," *Beinn Bhreagh Recorder*, September 30, 1909, p. 255, https://memory.loc.gov/mss/magbell/307/30700101/30700101.pdf.
63. KEB [Mrs Baldwin], *Beinn Bhreagh Recorder*, September 26, 1909, p. 231, https://memory.loc.gov/mss/magbell/307/30700101/30700101.pdf.
64. Alexander Graham Bell, "Third Flight Sept. 30," *Beinn Bhreagh Recorder*, September 30, 1909, p. 254, https://memory.loc.gov/mss/magbell/307/30700101/30700101.pdf.
65. Cecil R. Roseberry, *Glenn Curtiss, Pioneer of Flight* (New York: Syracuse University Press, 1991), 138.
66. Library of Congress, "3-2-12 Copied from the *Beinn Bhreagh Recorder* for April 14, 1910. The Flying-Machine of the Future; Aviation - 16 as Conceived in 1892 by Alexander Graham Bell," Alexander Graham Bell Family Papers, Library of Congress, https://memory.loc.gov/mss/magbell/375/37500701/37500701.pdf.
67. "Alexander Graham Bell: Kites, Flying Machines, and Hydrofoils," *Zooba American History*, accessed February 3, 2017, american_historyil.tripod.com/americanhistory/id9.htm.
68. Lawrence Surtees, "Bell, Alexander Graham," in *Dictionary of Canadian Biography*, vol. 15, www.biographi.ca/en/bio/bell_alexander_graham_15E.html.
69. "Bell, Alexander Graham," *National Aviation Hall of Fame*, 2017. http://www.nationalaviation.org/our-enshrinees/bell-alexander/.
70. Wayne Mutza, *Helicopter Gunships: Deadly Combat Weapon Systems* (Specialty Press, 2010), 7.
71. Robert V. Bruce, *Bell: Alexander Graham Bell and the Conquest of Solitude* (New York: Cornell University Press, 1973), 362.
72. National Academies, *Background Papers for a Drilling Technology Workshop, Park City, Utah, June, 1975*.
73. http://www.invaluable.com/auction-lot/bell,-alexander-graham.-extraordinary-historic-av-48-c-c32878efba.
74. Tim Cook, "The Great War in the Air," *The Canadian Encyclopedia Online*, August 1, 2014, www.thecanadianencyclopedia.ca/en/article/the-air-war.

# Additional Sources

Peter Busby, *First to Fly: How Wilbur & Orville Wright Invented the Airplane* (New York: Crown, 2003).

H. Gordon Green, *The Silver Dart: The Story of J.A.D. McCurdy, Canada's First Pilot and the First Airplane Flight in the British Empire* (Sydney, NS: Breton Books, 2014).

Canada at War blog, https://canadaatwarblog.wordpress.com/.

Hydroplane History website, http://www.lesliefield.com/other_history/alexander_graham_bell_and_the_hydrofoils.htm.

Wind Climbers Kite Club website, www.windclimbers.ca/alexander-g-bell.

Wright Brothers Aeroplane Company website, http://www.wright-brothers.org.

# Acknowledgements

I was challenged in a writing class taught by Paul MacDougall to write about things that are not well known by most people. That short story turned into this book, which Mabel Bell wanted to credit her husband's work in the development of aviation. The author of *Distinction Earned*, Paul MacDougall is an award-winning writer of fiction, drama and non-fiction who lives in Sydney, Nova Scotia. He teaches at the Cape Breton University. Thank you, Paul, for the inspiration to write this story on Cape Breton aviation history. My granddaughter Autumn's short life also inspired me to preserve memories of the past.

I would like to thank James Lorimer of Formac for his vision on this story, the entire staff at Formac for all of their assistance and Ashley Boyd for her advanced computer skills. Thank you to Madeline Harvey and Valerie Mason of the Bell Museum in Baddeck and Sara Grosvenor, President of the Alexander and Mabel Bell Legacy Foundation, for their work on photos.

# Index

Photographic references appear in **bold**.

Aerial Experiment Association (AEA)
  *Beinn Bhreagh Recorder* / bulletins, 32, 50, 74, 83
  charter ratification, 20
  formation, 17–26
  historical importance, 26–27, 40, 55, 93*n*35
  research, 33–34
  test program, 26, 31, 33, 36, 53
aerofoils, 32–33
ailerons, 38–41, 59, 60, 62, 71, 85, 93*n*35
anhedral wings, 60–62
aviation
  Bell's interest in, 10, 81
  competition, 18, 32–33, *see also* Wright brothers
  terms, 22, 32–34
Baddeck
  aviation history, 6–7, 54–58
  Bells' life in, 10
  community involvement, 10–11, 12, 13–15, 31, 54–58, 68, 70, 71–72, 74
  No. 1 (aeroplane), 68–72, 73, 89, 90, 91
  No. 2 (aeroplane), 52, 68–69, 71–74, 76, 77–78, 80, 84, 85, 86, **90**, 91
Baldwin, F. W. "Casey"
  AEA involvement, 19–21, **28**, 35, 37, 39–40, 47–50, 52, 62, 65–66
  CAC involvement, 68–70, 72–73, 76–79, 84–88, 90
  as chief engineer, 21–**22**, 25
  insight, 82, 83
  later acclaim, 67
Baldwin, Capt. Thomas, 20, 23
Beinn Bhreagh, 6, **7**, 10–11, 12, 13–**14**, 16, 17, 25, **21**, 22–25, 31, 47, 52, 55, 58, 65, 67–69, 72, **89**
Bell, Alexander Graham
  childhood, 9–10
  early work, 11–16, 19
  immigration to Canada, 9
  insight, 5, 11, 27, 62–63, 70, 73, 81, 87, 92
  personal writing, 10, 12–13, 16, 58, 75

Bell, Mabel, 5, **6**, 30
  assistance of, 13, 17, 18–19, 20, 65
  business acumen, 9
  letters, 56, 63, 65–66
Bell Museum, 32, 92
Bell's Law, 75, 77
Bentinck Farm, 72–78, 80, 92
biplane, 55–58
  Chanute–Herring, 32, 37
  designs, 33–36, 40, 52, 65–66, 71, 72, 85, 87
Bras d'Or Lake, 1, 7, 10, 11, 29–30, 38, 50, 56, 57, 87
Canadian government
  aeroplane sales, 68, 88–92
  Petawawa tests, 58, 68, 69–70, 71, 84, 89–90
Canadian Aerodrome Company (CAC)
  test program, 69, 84, 86
  historical importance, 84
  innovation, 71, 81–84
canard design, 34, 36–37, 41, 54, 59, 85
Cayley, George, 17, 26, 60
Chanute, Octave, 14–15, 21, 33, 34, 39, 64, 93*n*35
Chanute–Herring biplane *see* biplane
competition, 18, 32, *see also* Wright brothers, competition
Curtiss, Glenn
  AEA involvement, 19, 22, 23–**24**, 29, 33, 40, 41–46, 52, 55, 64–66
  engines, 23, 24–**25**, 27, 37, 46, 47, 51, 54, 60, 85
  success, 64–66, 93*n*35
  Wright brothers, 24–25
Cygnet I (kite), 12, 13, 26, 27–31, 54, 86
Cygnet II (kite), 27, 47, 50, 54
Cygnet III (kite), 27, 87
death
  aviation pioneers, 10
  Selfridge's, 62–63, 77
dihedral wings, 60–62
dirigibles, 20, 23
Dhonnas Beag (hydrofoil craft), **48**, 49–50, **83**
"Dolly" *see* Macleod, Dolena
Fairchild, Daisy, 44–45, 56

firsts
  British Empire, 55, 57, 68, 75, 78–80, 84
  Canada, 68, 85, 87, 90
Fort Myer tragedy, 62–63, 76–77
Frost King (kite), 14, **15**, 30, 31
Grey, Governor General Earl, 88, 91
Hammondsport, **22**, 23, 24, **25**, 31–34, **35**, 36–37, 42, 44, 45, 47, 51, 53, 56, 58, 65
hang gliders, 33
Hargrave, Lawrence, 11, 15, 16, 18, 46
HD hydrofoil boats (Nos. 1–4), 52
Hubbard (monoplane) *see* Mike, The
hydrodromes, 22
hydrofoils, 48–49, 52, 84
hydroplane, 22, 47–52, 94*n*43
Jumbo (kite), 11
June Bug (aeroplane), 41–42, 44–46, 51, 62, **65**, 75–76
kite house, 11, 12, **69**, **70**, **71**
kites *see* tetrahedral kites
Langley, Samuel Pierpoint
  Bell's friendship, 15–16, 22
  inventions, 15, 16, 17, 21, 55, 64
Lilienthal, Otto, 15, 21, 32, 34, 36
Loon (seaplane), 51–52
MacLeod, Dolena "Dolly," 75–80
MacRae Nicholson, Mabel Eleanor, 75–77
Maunsell, Major G. S., 88, 90–91
McCurdy, Arthur Williams, 13–14, 19
McCurdy, J. A. Douglas, 13, 79
  AEA involvement, 8, 19–20, 25, 35, 44, 47, 52, 55–58, 63, 65, 75
  CAC involvement, 68–70–72, 74, 77–78, 84–90, 92
  as a "doer," 20
  insight, 70, 83–84
  politics and corporate work, 66–67
Mike, The (monoplane), 84, 86–87
Militia Council (Canada), 89, 91
monoplane, 33, 84, 86–87
Oionus kites, 50, 51, 85–86

patents, 20
  aeroplane, 33, 39, 43, 45–46, 62
  telephone, 8, 9
  tetrahedral kite, 13
Petawawa *see* Canadian government
Pratt Truss design, 34, 37, 61
Red Wing (aeroplane), 36–38, 40, 61
Rutherford, Col. R. W., 89–91
publicity, 17, 31, 58, 73, 74, 78–79
safety, concern for, 10, 11–12, 17, 26, 34, 49, 55, 56, 63, 75
*Scientific American* contest, 42–45
Secretary Aero Club of America, 42–44
Selfridge, Lt. Thomas E., 22–**23**
  AEA involvement, 19, 22–25, 27, **29**–31, 33, 36–37, 47
  death, 62–63, 77
  Wright brothers, 23, 62–63, 76
Silver Dart, 7, 8, 22, 47, 53, 59–63, 69, 70, 76, 78–79, 85, 87, 92
  flight 8, 53–58, 68, 90
telephone, 8, 58
tetrahedral kites, 23, 85
  construction, 11–13, 21, 47, 54, 60, 87
  flight, 11, **15**, **17**, 26, 27–**30**, 50
  patenting, 13
triplane, 85
Ugly Duckling (boat), **24**, 72, **74**
ultralight trike, 12, 87
White Kite (kite), 50
White Oionus (kite) *see* Oionus kites
White Wing (aeroplane), 22–**23**, **38**, 40–41, 46
wing doping, 41
wing warping, 39–40, 59, 93*n*35
Wright brothers, 4, 15, 21, 23, 93*n*35
  competition, 7, 24, 26, 33–34, 39, 41–46, 58, 59–63, 68, 76, 77